Letts

Revise IGCSE

**Complete Study
&
Revision Guide**

Graham Booth

Physics

FOUNDATION® BOOKS

Contents

General physics

Thermal physics

Properties of waves

Electricity and magnetism

Atomic and particle physics

The Earth and beyond

Preparing for the examination

Planning your study

The final three months before taking your IGCSE examination are very important in achieving your best grade. However, your success can be assisted by an organised approach throughout the course.

- After completing a topic in school or college, go through the topic again in Letts IGCSE Physics Guide. Copy the main points onto a sheet of paper or use a highlighter pen to emphasise them.
- A couple of days later try to write out these key points from memory. Check differences between what you wrote originally and what you wrote later.
- If you have written your notes on a piece of paper, keep this for revision later.
- Try some questions from the book and check your answers.
- Decide whether you have mastered the topic fully and make a note of any weaknesses you think you have. Concentrate on rectifying these weaknesses using your Letts IGCSE Physics Guide.

Preparing a revision programme

Atleast three months before the final examination, go through the list of topics in your Awarding Body's specification. Identify which topics you feel you need to concentrate on. It is a temptation at this time to spend valuable revision time on the things you already know and can do. It makes you feel good but does not move you forward.

When you feel you have mastered all the topics, spend time answering practice questions. Each time check your answers with the answers given. In the final couple of weeks, go back to your summary sheets (or highlighting in the book).

How this book will help you

Letts IGCSE Physics Guide will help you because:

- it contains the essential content for your IGCSE course without the extra material that will not be examined
- it contains Progress checks and IGCSE questions to help you to confirm your understanding
- it gives sample IGCSE questions with answers and advice from an examiner on how to improve
- the summary table will give you a quick reference to the requirements for your examination
- marginal comments and highlighted key points will draw to your attention important things you might otherwise miss.

improve your grade

1. Read the question carefully

Many students fail to answer the actual question set. Perhaps they misread the question or answer a similar question they have seen before. Read the question once all the way through and then again more slowly. Some students underline or highlight key words in the question as they read it through. Questions at IGCSE contain a lot of information. You should be concerned if you are not using the information in your answer.

2. Give enough detail

If a part of a question is worth three marks, you should make atleast three separate points. Be careful that you do not make the same point three times. Approximately 25% of the marks on your final examination papers are awarded for questions requiring longer answers.

3. Quality of Written Communication (QWC)

From 2003 some marks on IGCSE papers are given for the quality of your written communication. This includes correct sentence structures, correct sequencing of events and use of scientific words.
Read your answer through slowly before moving on to the next part.

4. Correct use of scientific language

There is important scientific vocabulary you should use. Try to use the correct scientific terms in your answers and spell them correctly. The way in which language is used is often a difference between successful and unsuccessful candidates. As you revise make a list of physical terms you meet and check that you understand the meanings of these words.

5. Show your working

All Physics papers include calculations. Many of these rely on physical relation. You should always state the relations and show your working in full. Then if you make an arithmetical mistake, you may still receive marks for correct science. However, you will not be awarded any marks if you cannot recall the correct relationship. Check that your answer is given to the correct number of significant figures and give the correct unit.

The following topics are covered in this section:

- Length, time, density
- Speed, velocity and acceleration
- Movement and force
- The effects of forces
- More about forces and their effects
- Work, efficiency and power

1.1 Length, time, density

LEARNING SUMMARY

After studying this section you should be able to:

- recall the basic units for measurement of area and volume
- convert the units of measurement from higher to lower order units and vice versa
- identify vernier caliper and screw gauge as instruments to measure small dimensions
- calculate the density of a solid.

Length is measured in metres; for large lengths or distances, we use multiples while for short lengths, we use sub-multiples.

1m = 10 dm = 100 cm = 1000 mm

Submultiples: 1 dm = 1/10 m, 1 cm = 1/10 dm = 1/100 m

1 km = 10 hecto m = 100 deca m = 1000 m

1 mm = 1/10 cm = 1/100 decim = 1/1000 m

KEY POINT

Remember:

1 kilo	= 1000 times		1 deci	= 1/10th part

Remember:
1 kilo = 1000 times   1 deci = 1/10th part
1 hecto = 100 times   1 centi = 1/100th part
1 deca = 10 times   1 milli = 1/1000th part

Mass is measured in kilograms, kg.

1,000 grams = 1 kilogram;   1 milligram = 1/1000 g.

Time is measured in second, s.

Here is a full table of important prefixes.

Prefix	Abbreviation	Multiples	Example
atto	a	10^{-18}	1 attometre = 10^{-18} m
femto	f	10^{-15}	1 femtometre = 10^{-15} m
pico	p	10^{-12}	1 picogram = 10^{-12} g
nano	n	10^{-9}	1 nanosecond = 10^{-9} s
micro	μ	10^{-6}	1 micrometre = 10^{-6} m
milli	m	10^{-3}	1 milligram = 10^{-3} g
centi	c	10^{-2}	1 centimetre = 10^{-2} m
deci	d	10^{-1}	
deca	d	10^{1}	
hecto	h	10^{2}	
kilo	k	10^{3}	
mega	M	10^{6}	1 megametre = 10^{6} m
giga	G	10^{9}	
tera	T	10^{12}	1 terasecond = 10^{12} s
peta	P	10^{15}	
exa	E	10^{18}	

The **area** of an object or space is found by multiplying its length with its width. Since we are multiplying two quantities together that are both measured in metres, the unit of area is square metre (or metre squared) written as m².

If you look at a square of side 1 metre, its area must be 1m x 1m = 1m².

Note each side of the square is 1m long, or 100 cm. So in terms of square centimetres, 1 metre square has an area of 100 cm x 100 cm = 10,000 cm² **Fig. 1.1**.

Fig. 1.1

EXAMPLE

How many mm² are there in 1 m²?

Ans: One million, or 10⁶

The **volume** of an object is the amount of space it occupies. The volume of a cube is found by multiplying its length, breadth and height together. Since all these have the same value, the volume is found by cubing the length of a side.

So a cube with a side length of 1m has a surface area of 6 m² (there are six sides), but a volume of 1 m³.

EXAMPLE

How many (a) cubic centimetres, (b) cubic millimetres are there in 1 m³?

Ans: (a) 10⁶, (b) 10⁹

Measuring length and time

The method used for measuring the length of an object depends upon the size and the situation. For regular objects of normal size, a ruler or metre stick is used. For measurment of dimensions of objects, less than a centimetre, we use vernier calipers or micrometer screw guage.

For even smaller lengths, we can use a travelling microscope.

Fig. 1.2 Vernier calipers

Fig. 1.3 Micrometer screw gauge

For measurement of astronomical or interstellar distances, say the distance to Sirius – one of the brightest stars in the night sky, normal units of length are not enough. Here, we use the light-year or the distance travelled by light in one year. As you know that light travels at almost 300 million metres every second, you may get an idea of just how large a distance a light-year is!

Sometimes, we may be required to find the volume of an irregular object, such as a stone. Clearly, measuring with a ruler is going to lead to all sorts of errors. The method we use is to immerse the stone in a graduated cylinder which contains some water, so that we can note the rise in the water level and hence read off the change in volume. For wide objects, that do not fit in a graduated cylinder, we can use an overflow can where the object is immersed after filling the can to its spout with water. The water which overflows is collected in a graduated cylinder and the volume of the stone is found, as it is equal to the volume of displaced water.

Measurement of time

Time is measured in seconds. Up to the sixteenth century, time was measured through water-clocks, sand-timers, graduated candles, etc., all based on the length of the day as observed by the astronomers. With the coming of machinery, clock work mechanisms became increasingly accurate, and the hour, minute and second became uniform in many countries. Today's atomic clocks are exceedingly accurate, and a second is defined accurately in terms of specified atomic oscillations.

> A shoe box has a length of 30 cm, a width of 25 cm and a depth of 20 cm.
>
> **EXAMPLE** Find its (a) surface area in m² and (b) its volume in m³.
>
> **Ans**: (a) 2(750 + 600 + 500) cm² = 3700 cm² = 0.37 m²
>
> (b) 30 x 25 x 20 = 15,000 cm³ = 0.015 m³

Density

The density of a substance is defined as its mass per unit volume. It is often represented by the Greek letter rho (ρ).

$$\rho = \frac{m}{v}$$

> It is useful to remember that the density of water is 1 g/cm³ or 1000 kg/m³, and that the density of air = 1.29 kg/m³.

The units of density are kg/m³. When dealing with small objects, the unit of gcm⁻³ is often used (a sugar cube has a volume of approximately 1 cm³).

If something is less dense than water, it floats. Wood, for example, has an average density of 700 kg/m³, which is less than that of water. Ice is just slightly less dense than water, which is why an iceberg floats, although most of it lies beneath the surface of the ocean.

Gold, on the other hand, has a very high density and hence it sinks.

> **EXAMPLE**
>
> Calculate the density of a metal, given that a cube of side 3 cm, has a mass of 216 g.
>
> **Ans**: The volume of metal = 3 x 3 x 3 = 27 cm³. Since density = mass/volume , we calculate the density of the metal as 216 ÷ 27 = 8 g/cm³.

Experiment to measure the density of a stone

- First find the mass of the stone
- fill an overflow can with water and allow the excess water to drain away
- gently immerse the stone in the can, collecting all the water that overflows in a clean graduated cylinder
- read the volume of water that has been collected in the graduated cylinder, this equals the volume of the stone
- knowing the stone's mass and its volume, calculate its density.

stone

measuring cylinder

Fig. 1.4

> **EXAMPLE**
>
> (a) Why was the mass of the stone found before immersion and not afterwards?
>
> (b) Why did the instructions specify that the graduated cylinder had to be clean?
>
> **Ans**: (a) If mass were found after immersion, the stone would be wet and the mass registered would be slightly greater, thus introducing unnecessary error.
>
> (b) The dust accumulated in a dirty graduated cylinder would cause the volume reading to be increased, again introducing error.

1.2 Distance, speed, velocity and acceleration

LEARNING SUMMARY

After studying this section you should be able to:

● *calculate the speed of an object from a distance–time graph*
● *recall and use the relationship between acceleration, change in velocity and the time taken*
● *interpret velocity–time graphs.*

Distance–time graphs

If you go on a journey, the **distance** that you have travelled can only stay the same or increase, so a **distance–time graph** for the journey cannot have a negative slope or **gradient**. The graph, **Fig. 1.5**, shows a distance–time graph for a cycle ride.

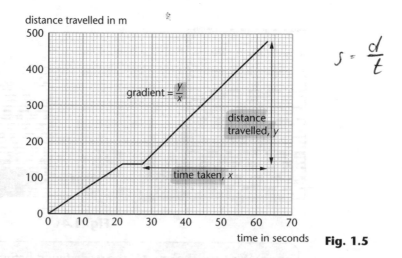

distance travelled in m

gradient = $\frac{y}{x}$

distance travelled, y

time taken, x

time in seconds

$$s = \frac{d}{t}$$

Fig. 1.5

The gradient of the graph line gives information about the cycle ride:

> **Where the graph line is horizontal, its gradient is zero.**

● between 22 s and 27 s the gradient is zero, showing that the cyclist was not moving

● calculating the **speed** of the cyclist using *speed = distance travelled ÷ time taken* is equivalent to calculating the gradient of the graph

● the graph has a steeper gradient between 27 s and 63 s than between 0 s and 22 s, showing a greater speed.

Speed, displacement and velocity

○ velocity
→ VECTOR

Knowledge of the speed of an object only tells you how fast it is moving, but its **velocity** also gives information about the **direction** of travel. The direction may be specific, as is the case when describing the velocity of an aircraft, or it may be relative, where velocity in one direction is described as positive and that in the opposite direction is negative.

> **Air traffic controllers tell pilots what velocity to fly at. The direction is given in terms of the points of the compass.**

> **KEY POINT**
> It follows that a speed–time graph can only have positive values, but a velocity–time graph can have both positive and negative values, representing motion in opposite directions.

A **displacement–time graph** gives more information than a distance–time graph:

- displacement is the distance an object moves from a fixed position, so it can decrease as well as increase
- displacement can have positive and negative values to show movement in opposite directions
- the gradient of a displacement–time graph represents velocity; a negative gradient represents movement in the opposite direction to that represented by a positive gradient.

Acceleration and graphs

Acceleration involves a **change in velocity.** Speeding up, slowing down and changing direction are all examples of acceleration.

> **A negative acceleration in the direction of motion is sometimes called a deceleration; it represents a decrease in speed.**

> **KEY POINT**
> Acceleration is the change in velocity per second. It is calculated using the relationship:
> acceleration = change in velocity ÷ time taken
>
> $$a = \frac{v-u}{t}$$
> where a is the acceleration of an object whose velocity changes from u to v in time t.
> Acceleration is measured in m/s^2.

The diagram, **Fig. 1.6**, shows a speed–time graph for part of a car journey.

speed in m/s

increase in speed, y

gradient =

time, x

time in s

$$a = \frac{(v-u)}{t}$$

Fig. 1.6

In this graph:

- the speed can only have positive values
- if the direction of motion does not change the acceleration can be calculated as *acceleration = change in speed ÷ time taken*; this is equivalent to calculating the gradient of the graph
- the steeper the gradient, the greater the acceleration that it represents
- a negative gradient shows a decrease in speed.

> If the direction does not change, the change in speed is the same as the change in velocity.

> **KEY POINT** The gradient of a speed–time graph represents the acceleration in the direction of motion.

The graph also gives information about the distance that an object travels:

- distance travelled = average speed × time
- the distance travelled in the first 10 s is equal to $\frac{1}{2} \times 14.5 \times 10 = 72.5$ m; this is represented by the shaded area on the graph
- similarly, the distance travelled during the next 25 s is represented by the area of the rectangle between the graph line and the time axis.

> The formula for distance travelled is derived from that for calculating speed.

> **KEY POINT** On a speed–time graph, the area between the graph line and the time axis represents the distance travelled.

A velocity–time graph differs from a speed–time graph because velocity can have both negative and positive values, showing motion in opposite directions.

On a velocity–time graph:

- if the gradient and velocity both have the same sign, the object is accelerating in the direction of motion
- if the gradient and velocity have opposite signs, the object is decelerating in the direction of motion
- the total area between the graph line and the time axis represents the distance travelled.

> When working out the distance travelled, areas below the time axis count as positive; they are simply added on to areas above the axis.

(handwritten note in margin)
∘ gradient + velocity
= same sign
→ accelerating in direction of motion

> **PROGRESS CHECK**
> 1. Calculate the speed of the motion shown during the first 22 s in **Fig. 1.5**.
> 2. Calculate the acceleration of the motion shown in the first 10 s in **Fig. 1.6**.
> 3. Calculate the distance travelled during the 50 s of motion shown in **Fig. 1.6**.
>
> 1. 6.4 m/s; 2. 1.45 m/s²; 3. 607.5 m.

Earlier, we said that velocity was different from speed, because velocity not only told us how fast an object is moving, but also the direction in which it is moving.

So, if we said that an aeroplane was flying at 400 km/h, we would be stating its speed. If we stated that the plane was flying at 400 km/h due North, we would be stating its velocity.

Speed, which has only magnitude (or size),,is a scalar while velocity is a vector.

Distance has only magnitude and hence it is a scalar while displacement is a vector.

Acceleration is a vector.

Here are some more examples:

SCALAR	VECTOR
Distance	Displacement
Speed	Velocity
Mass	Acceleration
Density	Force

Graphical/geometrical representation of a vector

A vector is represented by a straight line with an arrow-head at one of its ends. The length of the straight line represents the magnitude of the vector and the arrow-head represents its direction.

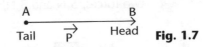

Tail $\overrightarrow{P}$ Head **Fig. 1.7**

Point A is called **tail** of the vector and point B is called the **head**.

Composition (addition) of vectors

The process of determining the resultant vector of two or more component vectors is known as the composition of vectors.

Adding physical quantities

What is the result of adding 2 kg to 2 kg? There is only one possible answer to this question, 4 kg. But if the question is changed slightly to "what is the result of adding together two 2 N forces?" the answer could be any value between 0 N and 4 N. This is because, unlike mass, forces have a **direction**. Quantities that have size only are called **scalars**, but those that also have a direction are **vectors**, and different rules of addition apply. The table gives some examples of vector and scalar quantities.

The sum of two vector quantities is sometimes called the resultant.

vectors	scalars
force	mass
displacement	distance
velocity	speed
acceleration	temperature

If two vector quantities of the same type act in parallel on the same object, working out their effect is a case of simple addition or subtraction. This is shown in the diagram, **Fig. 1.8**.

Fig. 1.8

To add together two vector quantities acting at right angles to each other, the following procedure is used:

● to a linear scale, draw an arrow that represents one of the vectors in size and direction

● starting at the end of the first arrow, draw a second arrow that represents the other vector

● a line drawn from the beginning of the first vector to the end of the second vector represents their sum, in size and direction.

> It does not matter which vector is drawn first, the result is the same if the vectors are drawn in either order.

The diagram, **Fig. 1.9**, shows the application of this method to find the sum of two forces, 5 N and 3 N, acting at right angles on the same object.

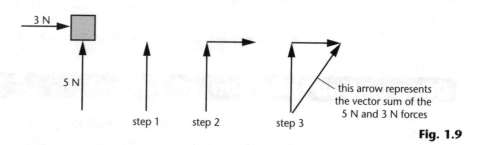

this arrow represents the vector sum of the 5 N and 3 N forces

step 1   step 2   step 3

Fig. 1.9

In this case the sum of these vectors is a force of 5.8 N acting at an angle of 31° to the 5 N force.

Methods of composition of vectors

There are three methods (of composition of vectors or **laws of vector addition**):

● **The Triangle law**

● **The Parallelogram law**

● **The Polygon law**

The Triangle law

If two unbalanced coplanar vectors acting simultaneously on a particle or at a point on a body are represented in magnitude and direction by the two sides of a triangle taken in the same order, then their resultant will be completely (in magnitude and direction) represented by the third side of the triangle.

Fig. 1.10 Vector representation

$$\vec{AB} + \vec{BC} = \vec{AC} \text{ or } \vec{V_1} + \vec{V_2} = \vec{R}$$

Parallelogram law

If two unbalanced coplanar vectors, acting simultaneously on a particle or at a point on a body, are represented in magnitude and direction by the two adjacent sides of a parallelogram drawn from a point, then their resultant will be completely (in magnitude and direction) represented by the diagonal of the parallelogram passing through the same point.

$$\vec{OA} + \vec{OB} = \vec{OC}$$
$$\text{or} \quad \vec{V_1} + \vec{V_2} = \vec{R}$$

Fig. 1.11 Vector representation

Motion graphs

Here are some graphs associated with motion:

1. Distance-time graph showing different velocities

2. Position-time graph for positive acceleration

3. Position-time graph for negative acceleration

4. Velocity of boy-1 is more than that of boy-2

5. Comparison of running and walking on position-time graph

6. Velocity-time graph for various accelerations

7. Velocity-time graph for negative acceleration

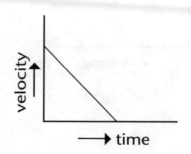

8. Velocity-time graph for positive acceleration

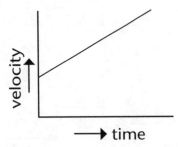

9. Velocity-time graph. PQ represents constant velocity, OB represents constant acceleration and AC represents infinite accleration

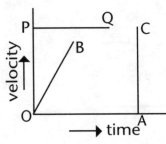

Some inferences from position-time and velocity-time graphs

- If the x - t graph is a straight line parallel to X-axis, then the body is at rest.

- The straight line inclined to time axis in x-t graph represents constant velocity.

- In x - t graph, the straight-line inclined to time axis at an angle greater than 90° shows negative velocity.

- No line in x - t graph can be perpendicular to time axis because it will represent infinite velocity.

- If the x - t graph is a curve whose slope decreases continuously with time, then the velocity of the body goes on decreasing continuously and the motion of the body is retarded.

- If the x - t graph is a curve whose slope increases continuously, then the velocity of the body is increasing continuously and the body is accelerated.

- If the v - t graph is a straight line parallel to the time-axis, then the acceleration of the body is zero.

- If the v-t graph is a straight line inclined to the time-axis with a positive slope, then that body is moving with constant acceleration.

- If the v - t graph is a straight line inclined to the time-axis with a negative slope, then the body is retarded.

- The velocity and acceleration of a body need not be in same direction.

> Always remember that the area under a velocity-time graph equals the distance travelled.

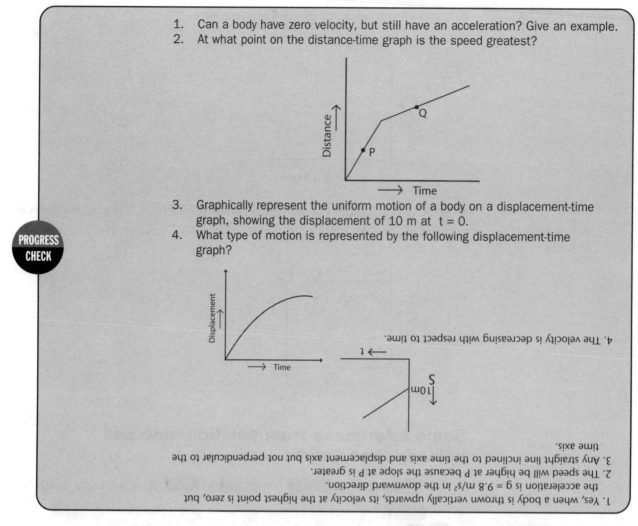

PROGRESS CHECK

1. Can a body have zero velocity, but still have an acceleration? Give an example.
2. At what point on the distance-time graph is the speed greatest?

3. Graphically represent the uniform motion of a body on a displacement-time graph, showing the displacement of 10 m at t = 0.
4. What type of motion is represented by the following displacement-time graph?

4. The velocity is decreasing with respect to time.

3. Any straight line inclined to the time axis and displacement axis but not perpendicular to the time axis.

2. The speed will be higher at P because the slope at P is greater.

1. Yes, when a body is thrown vertically upwards, its velocity at the highest point is zero, but the acceleration is g = 9.8 m/s² in the downward direction.

1.3 Movement and force

LEARNING SUMMARY

After studying this section you should be able to:

● **describe how balanced and unbalanced forces affect the motion of an object**
● **recall and use the relationship between force, mass and acceleration**
● **explain how a falling object reaches a terminal velocity.**

Starting and stopping

A force is a push or pull of one body on another

Our everyday experience of motion tells us that things cannot keep moving without application of a force. Remove the force and the motion eventually stops. Prior to the work of Galileo and Newton, it was thought that there is only one force involved in motion – the **driving force**. Newton realised that there are "unseen" forces such as **friction** and **air resistance**. If you push a book across a table, it stops moving because the friction between the book and the table is an **unbalanced force**.

The size of the friction force depends on the roughness of the surfaces: the rougher the surfaces, the greater the friction force.

friction force
acting on book

movement

Fig. 1.12

Forces due to friction:

- oppose slipping and sliding

- always act in the opposite direction of any motion

- cause **heating** and **wearing** of surfaces that rub together.

> Friction between a shoe and the ground prevents it from slipping. The force that propels a person who is walking is the forwards push of the ground on the shoe.

Friction is essential for walking, as well as starting and stopping the motion of a bicycle, bus or train.

Fig. 1.13

When you set off on a bike:

- the **wheel** pushes **backwards** on the road

- forces always act in pairs, so if object A pushes or pulls object B, then B pushes or pulls A with an equal-sized force in the opposite direction

- the **road** pushes **forwards** on the wheel, causing the bike to move.

The forces between the wheel and the road are frictional forces; without friction the wheel would just spin round.

> High-friction road surfaces are often used at roundabouts and traffic lights to help vehicles to stop in wet or icy conditions.

Friction is also needed to stop the bike while braking. When the brakes are applied, friction between the brakes and the wheel causes the wheel to stop rotating. Since the wheel will slip forward along the road, friction, which always acts to oppose motion, pushes backward on the wheel; thus bringing it to a halt. If there is in sufficient friction, the bike skids as it slides along the road surface.

A question of balance

As you cycle along at a steady speed, resistive forces act. The main resistive force is **air resistance**. To maintain a constant speed in the same direction, the resistive forces need to be **balanced** by the driving force, so that equal-sized forces act both forwards and backwards. *in the direction*

Changing **speed** or **direction** requires an **unbalanced force** in the direction of the change:

- when speeding up the driving force is bigger than the resistive force

- when slowing down the resistive force is bigger than the driving force

- when turning a corner friction between the road and the wheels causes a sideways-force.

The diagram, **Fig. 1.14**, shows the **balance** of forces acting on a cyclist when speeding up, travelling at constant speed and braking.

When braking, cyclists usually stop pedalling. This is why a driving force is not shown in the right hand diagram.

Fig. 1.14

Force and acceleration

When the forces acting on an object are balanced, there is no change in its motion. It either remains stationary or moves at a constant velocity, i.e. there is no change in speed or direction. An **unbalanced force** causes a **change in velocity** and the object accelerates.

ACCELERATION ← F
→ 3

The acceleration caused by an unbalanced force:

- acts in the direction of the unbalanced force
- is proportional to the size of the unbalanced force
- is inversely proportional to the mass of the object.

Two quantities are proportional if doubling the size of one causes the other to double. They are inversely proportional if doubling the size of one causes the other to halve.

$F = m \times a$

KEY POINT

The relationship between the size of an unbalanced force, *F*, and the acceleration, *a*, it causes when acting on a mass, *m*, is

force = mass × acceleration

$F = m \times a$.

This relationship is used to fix the size of the unit of force, the newton (N), as the force needed to accelerate a mass of 1 kg at a rate of 1 m/s².

It is important to remember that this relationship applies to the size of the **unbalanced** force acting, rather than any single force. The size of the unbalanced force due to two forces acting along the same line is their sum if they are in the same direction and their difference if they act in opposite directions. In the diagram, **Fig. 1.15**, the size of the unbalanced force is 1000 N in the **forwards** direction.

1000 N

2000 N

Fig. 1.15

Stopping distance

sd = N + S
= T + B

The distance that a vehicle travels between the driver noticing a hazard and the vehicle stopping is known as the **stopping distance**:

- stopping distance consists of **thinking distance** and **braking distance**
- thinking distance is the distance travelled during the driver's reaction time, the time between noticing the hazard and applying the brakes
- braking distance is the distance travelled while the vehicle is in the process of coming to rest.

The diagram, **Fig. 1.16**, shows how the stopping distance depends on vehicle speed.

30 mph

9 m   14 m   thinking distance   braking distance

60 mph

18 m   56 m

Fig. 1.16

> This diagram shows that thinking distance is proportional to vehicle speed and braking distance is proportional to the square of the vehicle speed.

These distances could be greater:

- if the driver is tired or affected by any drugs or alcohol; the reaction time and the thinking distance are increased
- if the road is wet or icy or the tyres or brakes are in poor condition; the braking distance is increased
- if the vehicle is fully loaded; the extra mass reduces the deceleration during braking, so the braking distance is increased.

> Some medicines contain drugs that can make the user drowsy. They carry a warning that a person should not drive after taking the medicine.

Stopping quickly

In a head-on collision, vehicles can be brought to a halt in a very short time. The driver and passengers in a vehicle would **carry on moving** until they hit something that could cause sufficient force to decelerate them. This would probably be the car windscreen or the seat in front, and in either case the force on impact could cause severe injury.

> Steering columns can also cause injury; in modern cars they are designed to collapse on impact.

Seat belts reduce the injury to drivers and passengers:

- seat belts prevent the occupants of a vehicle from hitting solid objects in front   *ELASTIC*
- they stretch slightly so that the wearer is brought to rest over a **longer time** than by hitting the object in front
- this reduces the **deceleration** of the wearer and the force required to stop him.

crumple zone   **Fig. 1.17**

crumple zone   rigid passenger cage

Modern cars and trains are also designed with **crumple zones** at the front and rear. The vehicle takes longer to stop in a collision, reducing the deceleration and the force required to stop the passengers.

Bodies in free fall

All bodies in the universe exert an attractive force called gravity – the bigger the body, the greater the force. We live on the surface of the Earth, which is a sizeable planet. It exerts a force on us, pulling us to its surface. If something drops from our hand, it falls to the surface of the Earth under the force of gravity. If we were on the moon and something fell from our hand, it would fall to the surface of the moon but at a slower rate because the moon, being a smaller and less dense body, exerts a smaller force of gravity. Now, as we will see, when a force acts on a body– the body accelerates falling with greater and greater velocity. On the Earth, in the absence of air, the body would continue to accelerate right upto the time when it strikes the Earth. The acceleration-time graph is a horizontal straight line while the velocity-time graph is a straight line with positive slope that passes through the origin.

Without air resistance, all objects would fall at the same rate. If a hammer and a feather were dropped together on the moon (where there is no air), they would strike the ground at the same time. On the Earth, the feather floats gently to earth, being buoyed up by the air.

Take a sheet of paper and let it fall, watch it gently move from side to side as it falls. Now crumple the same sheet into a ball and watch how it falls directly to the ground.

Analysis of air resistance is a complex exercise. Clearly, it depends on the shape of the object, among other factors. It also increases as speed increases. Car and plane designers spend a lot of time analysing results obtained from wind tunnels and trying to reduce air friction (or drag) as much as possible in order to produce a machine which is more fuel efficient and gives better performance. But for all objects falling to Earth, their initial acceleration (which has the value of 9.8 m/s^2) is reduced as its speed increases. Ultimately, the downward acceleration of the object is zero, when air resistance or upthrust balances the force of gravity. The object then falls with constant speed, or **terminal velocity**. You should know the shape of the speed-time graphs for a body falling to the Earth (a) without air resistance and (b) with air resistance.

(a) no air resistance   **(b)** with air resistance

Fig. 1.18 Graph for a freely falling body

1.4 The effects of forces

LEARNING SUMMARY

After studying this section you should be able to:

- *describe the behaviour of a spring and a rubber band when subjected to an increasing stretching force*
- *apply the principle of moments to calculate the forces acting on a stable structure.*

Weight of a body

$W = m \times g$

The force with which the Earth attracts a body is called the weight of body on the earth. It is a vector, which has both magnitude and direction. On Earth, the weight is always directed to the centre of the Earth (which is a sphere). Like all forces, weight is measured in **newtons**.

Mass of a body

The mass of a body is a measure of the quantity of matter contained in it. It is a scalar quantity. Its unit is **kg**. It is measured by a physical or chemical balance.

KEY POINT

The weight of a body is not constant. It changes from place to place on the Earth and also with the height from the surface of the earth and depth below the surface of the Earth. However, the mass of a body is always constant.

Let us take the example of a bag of sugar which is usually sold in units of 1 kilogram. The mass of sugar is 1 kg. Now the Earth's gravity is strong enough to pull each kilogram of mass downward with a force of 1 **newton**. We say that the Earth's gravitational field strength, g = 1 N/kg.

The moon has a smaller gravitational field strength, equal to 1.6 Nkg-1

So our bag of sugar of mass 1 kg has a weight of 9.8 N on the Earth, while on the moon, it has the same mass of 1 kg (the amount of sugar is still the same as long as we haven't used any!) but its weight is only 1.6 N.

On the giant planet Jupiter (where the field strength is very high), our bag of sugar still has a mass of only 1 kg, but it would be far too heavy to lift as Jupiter's gravitational field strength is so high that it would pull it downwards with a force far greater than we could counteract.

Mass can also be described as that property of a body, which *resists change in motion*. After all, to push a car we require a great deal of force. This is because the mass of the car is huge and hence, its reluctance to move. (At least the car has wheels; if only the chasis without the wheels lay on the ground, it would take the combined force of many people to move it.

Forces and materials

Pulling or pushing on a material causes it to change its shape. Sometimes the change in shape is apparent, as in the stretching of a rubber band. On other occasions, such as when walking across a concrete floor, the change in shape is so minute that it is unnoticeable. Objects such as bridges are subjected to forces that compress some parts and stretch others. It is important that they are designed to withstand these changes in shape and not to undergo permanent deformation.

> When you walk across a floor or sit on a chair it compresses. This results in an upward force on you that balances your weight.

The diagrams, **Fig. 1.19**, show how the extension of a spring and a rubber band depends on the size of the stretching force.

In the case of the spring:

> The uniform pattern of extension of a spring is used in a spring balance or forcemeter, where equal increases in force result in equal increases in extension.

- the extension is proportional to the force up to the end of the straight line part of the graph, called the limit of proportionality (sometimes called the elastic limit)

- beyond this point the spring becomes harder to stretch, greater increases in force are required to cause the same increase in extension

- the spring may not return to its original size if stretched beyond the limit of proportionality.

Spring

Rubber band

Fig. 1.19

For the rubber band:

- the extension is not proportional to the stretching force, the band becomes easier, then more difficult to stretch as the force is increased

> This does not happen if the force is large enough to break the rubber band.

- the band always returns to its original size and shape when the stretching force is removed.

This property of stretching due to a force was first investigated properly by Robert Hooke. Hence, we have Hooke's law which states:

if a force is applied to a solid, the extension produced is proportional to the force applied upto the elastic limit.

Turning forces

The **turning effect** of a force is used for opening a door, riding a bike and even while flushing the toilet. All these examples involve objects that are free to **rotate** around a fixed point or **pivot**. The effect that a force has in causing rotation depends on:

- the **size** of the **force**
- the **distance** that it is **applied** from the **pivot**
- the **angle** at which the force is **applied**.

> **KEY POINT**
>
> The moment, or turning effect of a force, is calculated using the relationship:
> **moment = force × perpendicular distance to pivot**
> The moment of a force is measured in N m.

If the line of action of the force passes through the pivot, there is no turning effect.

When several turning forces act on an object, whether it turns round depends on the balance of the moments acting in an **anticlockwise** direction compared to those acting in a **clockwise** direction. If the **sum** of the moments in each direction is the **same** then the object is **balanced**.

The principle of moments applies to the moments calculated about any single pivot.

> **KEY POINT**
>
> The principle of moments states that, when a system is balanced:
> sum of clockwise moments = sum of anticlockwise moments.

Forces acting on a beam

When a heavy lorry travels across a beam such as a bridge across a road, the forces acting at the bridge supports change with the position of the lorry. The two rules used to calculate the size of these forces are:

- the **total** force acting **upwards** must equal the **total** force acting **downwards**
- the **clockwise** moment about each support must equal the **anticlockwise** moment.

The forces acting are shown in the diagram, **Fig. 1.20**. It is assumed that the weight of the bridge itself is small and can be ignored.

As the lorry travels from right to left, the force R_1 increases and R_2 decreases.

Fig. 1.20

Application of the above rules to this situation gives:

- $R_1 + R_2 = W$ since the sum of the upward forces equals the downward force
- $W × x_1 = R_2 × (x_1 + x_2)$, taking moments about the **left-hand** support
- $W × x_2 = R_1 × (x_1 + x_2)$, taking moments about the **right-hand** support.

For an object to be in equilibrium:

- no resultant external force must act on it (so it does not accelerate).

- no resultant moment acts on it (so it does not rotate).

Force, then, is a push or pull which moves (or tends to move) a body at rest or stops (or tends to stop) a moving body or changes its direction of motion. It is measured in newtons.

Centre of mass

If we take a metre stick and place our finger under its midpoint, the metre stick balances.

This is because we are supporting the ruler at its centre of mass. What is really happening is that the turning moment of the weight on the left-hand side of the finger balances the moment of the weight on the right-hand side. The net effect is that all its weight appears to act at its mid-point, where we are supporting it.

Every object has a centre of mass (C.M.). If the body is symmetrical, its C.M. will lie on the intersection of the lines of symmetry – the C.M. of a rectangle, for example, lies on the intersection of its diagonals and the C.M of a doughnut lies in the middle of the central hole.

Two points to note:

- if we balance an object on a point, its C.M. must lie directly above that point

- if we suspend an object from a point, its C.M. must lie directly beneath that point.

So to find the C.M. of an irregularly shaped object, say a piece of card, we suspend it from a point near its edge, along with a plumbline and mark the line made by the plumbline on the card. We repeat this process by suspending the card from different points near its perimeter.

> **Plumbline is a vertical line used as a reference to measure deflections.**

The lines thus marked should intersect at a point – this is the centre of mass.

Stability

All objects have a **centre of mass**, the point at which the force of weight is taken to act. For a solid object in the shape of a cuboid, the centre of mass is at the centre of the cuboid. For a human being it is in the region of the navel, and for a tyre it is in the space in the centre. The diagram, **Fig. 1.21**, shows the positions of the centres of mass of a traffic cone, a rubber ring and a person.

> **In the case of a non-uniform object such as a hammer, the centre of mass is close to the heavy end.**

Fig. 1.21

If an object is suspended freely, when it is at rest its centre of mass is directly below the point of suspension. In any other position the weight of the object has a **moment** (see section 1.4). The centre of mass of a shape in the form of a lamina can be found by:

> The moment has a turning effect that returns the object to its rest position.

- suspending the shape from a pin in a loose-fitting hole so that it is free to turn
- drawing a line vertically below the point of suspension when the object is at rest
- repeating this at other points
- the centre of mass is the single point where the lines cross.

This is shown in the diagram, **Fig. 1.22**.

pin

weight on string

Fig. 1.22

The centre of mass of a bowling pin is higher than that of a traffic cone and it has a much narrower base. These factors combine to make it topple over when it is tipped. When a traffic cone is tipped it returns to its base.

The reason for this is shown in the diagram, **Fig. 1.23**.

> If the cone had a wider base, more tilt would be needed to make it topple.

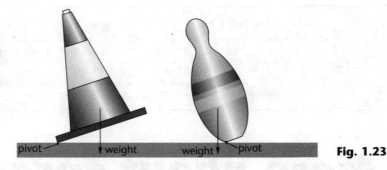

pivot   weight   weight   pivot   **Fig. 1.23**

The traffic cone is **stable** and the bowling pin is **unstable** because:
- when the cone is tipped its weight exerts a restoring moment
- when the pin is tipped the line of action of its weight falls outside the base, so the moment of the weight makes the pin fall over.

Note that if we want objects to be stable, we should keep the centre of mass as low as possible. This is why a ship carries ballast in its hull. The weight at the bottom lowers the centre of mass and hence keeps the ship stable.

Motion in a circle

Centripetal force

When the velocity of an object changes, we say that it is accelerating and a force must be acting on it. Since velocity is a vector, and has both magnitude and direction, an object may be accelerating even if the magnitude of its velocity (i.e. its speed) is constant but its direction is changing.

Think of a car travelling around a corner at a constant speed. You feel pushed against the side of the car – you feel a force – because the velocity of the car is changing. Its speed is constant, but it is accelerating. This acceleration is always directed towards the centre of the circular arc that the car is following. The force towards the centre is known as the centripetal force.

More examples of circular motion :

- **Bending of a cyclist** : A cyclist while going around a curve on a horizontal track has to bend himself a little from his vertical position, in order to avoid overturning. When he bends himself inwards, then a component of the reaction of the road provides him the necessary centripetal force for circular motion.

- **Whirling a stone**: A child whirling a stone on a piece of string feels that his hand is being tugged outwards by the stone, he needs to exert an inwards force with his hand to keep the stone moving in a circular path. If he lets go of the string, the stone will fly off at a tangent to the path.

- **The motion of the moon**: The moon orbits the Earth (actually, both Earth and moon orbit around their combined centre of mass, which lies beneath the surface of the Earth but not directly at its centre). It is the Earth's gravity that provides the circular force which keeps the moon in its orbit.

PROGRESS CHECK

1. What is the weight of a 10 kg mass on Earth?
2. Why does a rubber band not have a "limit of proportionality"?
3. Explain why door handles are mounted as far from the hinge as possible.

1. Weight = mg = 10 x 9.8 = 98 N; 2. The extension is not proportional to the stretching force; 3. To maximise the moment and minimise the force needed to open the door.

1.5 More about forces and their effects

LEARNING SUMMARY

After studying this section you should be able to:

- *use the equations of motion to solve problems*
- *describe the motion of a projectile*
- *explain how the principle of conservation of momentum applies to collisions*
- *explain how the stability of the object is affected by the position of its centre of mass*
- *explain why an object in circular motion is accelerating*
- *calculate the size of the unbalanced force required to maintain circular motion*
- *describe how Hooke's law applies to a spring*
- *explain the significance of the spring constant of a spring.*

Projectiles and momentum

Equations of motion

When an object moves at a constant speed, it travels equal distances in equal time intervals. An accelerating object travels increasing distances in successive equal time intervals. If the acceleration is constant, the relationships between the variables are described by the **equations of motion**.

> A constant acceleration is also known as a uniform acceleration.

The variables are:
- the initial velocity, u
- the final velocity, v
- the time of motion, t
- the displacement, s
- the acceleration, a.

> **KEY POINT**
> The equations of motion are:
> $v = u + at$
> $v^2 = u^2 + 2as$
> $s = ut + 1/2\ at^2$
> $s = 1/2\ (u + v)t$

> If a ball is thrown vertically into the air, when it returns to its starting point the displacement is zero, but it has travelled some distance.a

Each equation is a relationship between four variables, so if the values of three are known the other two can be calculated. When using these equations it is important to remember that:
- s represents displacement; this may not have the same value as the distance travelled
- v, u, a and s all have direction as well as size, so + and – signs should be used if, for example, the acceleration is in the opposite direction to the velocity.

Projectiles

When a ball is thrown horizontally, it starts to accelerate vertically as soon as it leaves the hand. If the effects of air resistance are negligible then its horizontal speed is constant. Different relationships apply to the horizontal and vertical motions and they should always be analysed independently.

> The common factor between the horizontal and vertical motion of a projectile is the time for which it is in the air.

> **KEY POINT**
> The equations of motion apply to the vertical motion.
> The relationship average speed = distance travelled + time taken applies to the horizontal motion.

Objects which are projected either horizontally or at the same angle to the horizontal are called projectiles. They all follow the same pattern of motion:
- horizontally, a projectile travels equal distances in equal time intervals
- vertically, a projectile travels increasing distances in successive equal time intervals.

The path of a projectile is shown in the diagram, **Fig. 1.24**. The steepness of the curve depends on the horizontal speed.

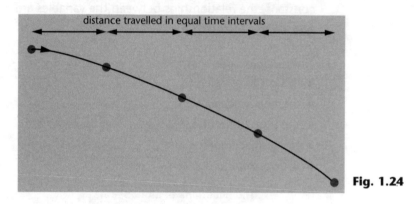
distance travelled in equal time intervals

Fig. 1.24

What is momentum?

Why is a heavy, slow-moving ball used in ten-pin bowling rather than a lighter, faster-moving ball? The heavier ball is more effective because it has more **momentum**.

> Momentum is a measure of how difficult it is to stop a moving object.

Momentum is a useful concept for predicting the outcome of a collision. The momentum of a moving object depends on:
- its mass
- its velocity.

KEY POINT

momentum = mass × velocity
$$p = m \times v$$
Momentum is a vector quantity and is measured in kg/ms or the equivalent unit N s.

> In a collision the momentum and total energy are always conserved.

When two objects collide there is no overall change in momentum, provided that the only forces acting are those between the objects themselves. The forces between the objects are equal in size and opposite in direction, and so are their changes in momentum. Momentum is always conserved – this is known as the **principle of conservation of momentum**.

Types of collision

When two "vehicles" collide on an air track the external resistive forces are small and so the principle of conservation of momentum can be used to predict the outcome. The fact that velocity and momentum have direction as well as size needs to be taken into account by using positive values for motion in one direction and negative values for motion in the opposite direction. The diagram, **Fig. 1.25**, shows an example of two "vehicles" colliding and sticking together.

before 1.5 m/s   0.3 m/s   after v = ?

0.4 kg   0.2 kg   0.6 kg

Fig. 1.25

The vehicles can be made to stick together by fitting one with a pin that sticks into some Plasticine on the other vehicle. A rebound collision occurs if the vehicles are fitted with repelling magnets.

If velocity to the right is taken to be positive then application of conservation of momentum gives:

0.4 kg × 1.5 m/s + 0.2 kg × –0.3 m/s = 0.6 kg × v

This gives a value for v equal to +0.9 m/s, the positive sign showing that the velocity is to the right.

In this type of collision the "vehicles" have less kinetic energy in total after they stick together than before they collide – some of it has been transferred to heat. When this occurs the collision is **inelastic**. In an **elastic collision** the total kinetic energy as well as the momentum is conserved.

When colliding objects stick together, the collision is always inelastic.

Conservation of momentum also applies to **jet** and **rocket** propulsion. A jet uses oxygen from the surrounding air to burn its fuel, but a rocket carries its own oxygen supply, so the engines can be used outside the Earth's atmosphere.

If a balloon is blown up and released without its neck being tied, the air inside gains momentum as it is forced through the neck. To satisfy conservation of momentum the balloon must gain the same amount of momentum in the opposite direction, so that the total momentum remains zero. A similar thing happens in rocket propulsion, shown in the diagram, **Fig. 1.26**.

exhaust gases gain momentum in this direction

rocket gains momentum in this direction

Fig. 1.26

As a rocket burns fuel, its acceleration increases due to its decreasing mass.

The exhaust gases gain momentum in the backward direction as they are forced out of the rocket. The rocket gains the same amount of momentum in the forward direction so that there is no overall change in momentum.

Vehicle safety

The effect of a force in changing the momentum of an object depends on:
- the size of the force
- the time for which it acts.

KEY POINT force = change in momentum ÷ time
$$F = \Delta p / t$$

This statement is equivalent to **force = mass × acceleration**. In a road accident, a vehicle can be brought to a halt very rapidly. If the driver and passengers had the same deceleration, they would experience large forces which could cause serious injury.

Injuries to the passengers in a vehicle are reduced by:

- **crumple zones** at the front and rear of the vehicle – these extend the time it takes for the vehicle to stop, reducing the deceleration of the vehicle
- **seat belts** – these restrain a passenger from colliding with parts of the vehicle, but stretch slightly to allow the passenger to carry on moving after the vehicle has stopped, reducing the deceleration of the passenger
- **air bags** – these inflate during a collision, reducing the pressure due to the retarding force acting on a passenger.

> When a vehicle stops, the passengers carry on moving until a retarding force acts on them.

PROGRESS CHECK

1. A bicycle travelling at 6 m/s increases its speed to 12 m/s in 12.0 s. Calculate its acceleration and the distance it travels while it is accelerating.
2. How does the horizontal motion of a projectile differ from its vertical motion?
3. Calculate the momentum of a 7 kg ten-pin bowling ball travelling at a speed of 2.5 m/s.

1. 0.5 m/s² and 108 m; 2. The horizontal motion is at constant speed. The vertical motion is accelerated; 3. 17.5 kg m/s.

Turning in a circle

Going round in circles

Some artificial satellites occupy circular orbits and others are in elliptical ones. A satellite in a circular orbit maintains a constant speed, while that of a satellite in an elliptical orbit varies in a similar way to that of a comet.

> **KEY POINT**
>
> For a satellite in a circular orbit:
>
> $$\text{orbital speed} = \frac{2\pi \times \text{orbital radius}}{\text{time period}}$$
>
> $$v = 2\pi r / T$$

Motion in a circle at a constant speed is accelerated motion. This is because the direction of motion, and therefore the velocity, is constantly changing. This acceleration is called **centripetal acceleration** and is directed towards the centre of the circle.

> An object moving in a circle has to keep moving towards the centre in order to stay the same distance away from it.

> **KEY POINT**
>
> The relationship between centripetal acceleration, orbital speed and the radius of the orbit is:
>
> $$\text{acceleration} = \frac{(\text{orbital speed})^2}{\text{radius}}$$
>
> $$a = v^2 / r$$

To whirl a rubber bung around in a circle, you have to keep pulling on the string. If you let go of the string, the bung moves in a straight line at a tangent to the circle.

All acceleration requires an unbalanced force and in the case of circular motion the force is called the **centripetal force**. The unbalanced force on a satellite is gravitational and that on an orbiting electron is due to electrostatic attraction. The diagram, **Fig. 1.27**, shows these forces.

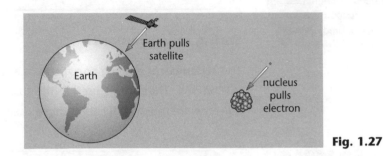

Earth pulls satellite

Earth

nucleus pulls electron

Fig. 1.27

It is important to remember that centripetal force is not a force that appears when an object moves in a circle. It is the unbalanced force that is acting on the object.

The size of the centripetal force on an object travelling in a circle:

● increases with increasing mass of the object
● increases with increasing speed of the object
● decreases with increasing radius of the circle.

KEY POINT

The relationship between the centripetal force and these variables is:

$$force = \frac{mass \times (orbital\ speed)^2}{radius}$$

$$F = mv^2/r$$

PROGRESS CHECK

1. Explain why a half-full milk bottle is more stable than a full one or an empty one.
2. A person with a mass of 65 kg travels in a circle of radius 8.5 m on a fairground ride. Calculate the centripetal force and centripetal acceleration when the speed of the person is 12.0 m/s.
3. Suggest why cars sometimes leave the road when going round a bend in icy conditions.

1. The half-full milk bottle has a lower centre of mass; 2. 16.9 m/s² and 1101 N; 3. The friction force is too small for the centripetal force required.

Some other effects of forces

Using springs

For small stretching forces, springs and metal wires are **elastic** – they return to their original size and shape when the stretching force is removed. If a spring or metal wire is stretched beyond the **elastic limit** then the behaviour is **inelastic** and permanent deformation takes place.

Fig. 1.28 shows the behaviour of a spring when subjected to an increasing stretching force. Metal wires show similar patterns of behaviour. Up to the limit of proportionality, springs and wires follow **Hooke's Law**.

Hooke's Law is very limited as it only applies to some materials over a limited range of forces. It cannot be used to make useful predictions.

 KEY POINT

Hooke's Law states that:
the extension is proportional to the force provided that the limit of proportionality is not exceeded.

Spring measurements

The **stiffness** or **spring constant** of a spring is a measure of how difficult it is to stretch it. The springs used in the suspension of a car and a washing machine are much stiffer than those used in retractable ballpoint pens.

A stiff spring has a larger value of spring constant than a weak one.

 KEY POINT

The spring constant of a spring, k, is defined by the equation:
force = spring constant × extension
$F = k \times x$

The diagram, **Fig. 1.28**, shows how to determine the spring constant as the gradient of a force–extension graph.

Some old clocks and watches use springs to store the energy needed to keep the mechanism in motion.

Springs **store energy** when they are stretched. When used in a car suspension, they absorb the energy of the vertical motion when the car goes over a bump in the road. The energy stored in a stretched spring is represented on a force-extension graph by the area between the graph line and the extension axis – this is the area shaded in the diagram above.

force

gradient = $\frac{y}{x}$ represents the spring constant

this area represents the energy stored at an extension x

extension

Fig. 1.28

 KEY POINT

If the extension of a spring or wire is linear:
energy stored = $\frac{1}{2}$ × force × extension
$E = \frac{1}{2} \times F \times x$

 PROGRESS CHECK

1. On a graph of force against extension for a metal wire, which part of the graph does Hooke's Law apply to?
2. A 15 N weight stretches a spring by 0.22 m.
   Calculate:
   (a) the spring constant.
   (b) the energy stored in the stretched spring.

1. The straight line portion;   2(a). 68 N/m; (b). 1.65 J.

1.6 Work, efficiency and power

LEARNING SUMMARY

After studying this section you should be able to:

● *describe everyday energy transfers*
● *recall and use the relationships for calculating work, power, kinetic energy and gravitational potential energy*
● *calculate the efficiency of an energy transfer.*

Work and energy transfer

Everything that happens involves **work** and **energy transfer**. Whenever a force makes something move **work** is being done and **energy** is being **transferred** between objects. The amount of work done and the amount of energy transferred are the same.

> The terms "work" and "energy transfer" have the same meaning.

KEY POINT

When a force, *F*, moves an object a distance *x* in its own direction:
work done = energy transferred = force × distance moved
$$E = F \times x$$
Work and energy are measured in joules (J) when the force is in N and the distance is in m.

Energy can be transferred and stored in a number of different ways:

> The term "potential energy" is often used as an abbreviated form of "gravitational potential energy".

● stored energy is **potential energy**; it can be **gravitational** due to position, **elastic** in a stretched spring or **chemical** in a lump of coal or a battery

● the energy stored in the atomic nucleus is **nuclear energy**; this is the energy source in a nuclear power station

● energy due to movement is **kinetic energy**; this includes the energy of a sound wave and the energy of particles in a gas

● the energy of an object due to its temperature is **thermal energy**; the thermal energy of a solid and a liquid comprises both kinetic and potential energy of the particles

> The term "heat" should not be used to describe infra-red radiation. Like all other forms of electromagnetic radiation, it has a heating effect when it is absorbed.

● **thermal energy** (heat) is transferred between objects by the processes of conduction, convection, evaporation and radiation

● energy is transferred by an **electric current** from a power supply to the components in a circuit

● energy is transferred between objects by electromagnetic radiation, or **radiant energy**.

Conversion of gravitational potential energy into kinetic energy:

When a body falls freely from a certain height, its potential energy starts decreasing and its kinetic energy starts increasing. However, the total mechanical energy remains conserved.

When a body starts to travel down an inclined plane, its potential energy decreases and it starts to gain kinetic energy.

If the inclined plane is smooth, then the potential energy of the body at the top of the plane converts wholly into kinetic energy at the bottom. However if the plane is rough, then some of the energy is converted into heat as work is done against friction. Hence, the kinetic energy at the bottom is less than the potential energy at the top of the plane.

Efficiency of energy transfer

In a coal-fired power station, for every 100 J of energy stored in the coal that is burned, only 40 J is transferred to electricity. **Energy**, like **mass** and **charge**, is a **conserved quantity**. This means that the total amount of energy remains the same, it cannot be created from or turned into a different quantity. In the case of the power station, 60% of the energy input is wasted and ends up as **heat** in the surrounding atmosphere and river.

This is shown in the energy flow diagram, **Fig. 1.29**.

The energy wasted as heat cannot be recovered as it only causes a small temperature rise in the surroundings.

Fig. 1.29

Gas-fired power stations are more **efficient**; they transfer more of the energy from the fuel to electricity. The **efficiency** of an energy transfer is the fraction or percentage of the energy input that is transferred to the desired output.

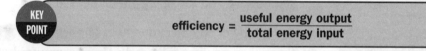

> **KEY POINT**
>
> $$\text{efficiency} = \frac{\text{useful energy output}}{\text{total energy input}}$$

Gravitational potential energy and kinetic energy

In the **pumped storage system** shown in the diagram, **Fig. 1.30**, surplus electricity generated at night when the demand is low is used to pump water from a low lake to a high one.

water released to
generate electricity
at peak demand

water pumped to
high level reservoir
at night using
surplus electricity

turbines and
pumps/generators

Fig. 1.30

The energy is stored as **gravitational potential energy**.

> **KEY POINT**
>
> **The change in gravitational potential energy when a mass, *m*, moves through a vertical height, *h*, is calculated using the relationship:**
> gravitational potential energy = mass × gravitational field
> strength × height
>
> **GPE = $m \times g \times h$**

A pumped storage system can supply electricity within two minutes after the opening of the valves. It takes several hours for a coal-burning power station to become operational.

At times of peak demand, the water is allowed to fall through vertical pipes, transferring energy from **gravitational potential energy** to **kinetic energy** as it does so. The **kinetic energy** of the moving water is transferred to the turbines. The turbines drive the generators in which the kinetic energy is transferred to electrical energy.

> **KEY POINT**
>
> **The kinetic energy of a mass, *m*, moving at speed *v*, is calculated using the relationship:**
> kinetic energy = $\frac{1}{2}$ × mass × speed²
>
> **KE = $\frac{1}{2} \times m \times v^2$**

The **power** output of the generator depends on its efficiency and the power input from the moving water. **Power** is the **rate of energy transfer**, the work done or the energy transferred each second.

At IGCSE, candidates often lose marks by confusing the units of energy and power.

> **KEY POINT**
>
> **Power is calculated using the relationship:**
> $$\text{power} = \frac{\text{work done or energy transfer}}{\text{time}}$$
> $$P = \frac{E}{t}$$
> **Power is measured in watts (W) when the energy is in J and the time is in s.**

All electrical appliances have a power rating. This is the electrical power input. The output power depends on the **efficiency** of the appliance. Appliances used for heating are much more efficient than those used for lighting and movement.

1. An electric motor is used to raise a lift cage through a height of 9.5 m in 6.0 s. The mass of the cage and its contents is 420 kg. The value of g is 10 N/kg.
   (a) How much work does the motor do?
   (b) Describe the energy transfer that takes place.
   (c) Calculate the output power of the motor.
   (d) The power supplied to the motor is 10 500 W. Calculate the efficiency of the motor.

(a) 39 900 J;   (b) electrical energy to gravitational potential energy and heat;   (c) 6 650 W;
(d) 0.63

Much research is now being done on alternative sources of energy. Oil and natural gas are being used at a rapid rate, coal is also being depleted though more slowly. Many third world countries are poor because trees are chopped down for fuel and building material – the damage is not just the loss of trees, but the loss of soil too that turns the landscape into a dustbowl unfit for agriculture. Large areas of several countries have been turned into arid deserts by the destruction of their natural vegetation.

Research into renewable sources of energy is being done by investigating:

- wind energy, where the combined power of many large windmills produce electricity

- wave and tidal energy where the movement of the Earth's oceans is harnessed

- geothermal energy, where the warmth of the underlying rocks is harvested
   ✓ expensive, difficult to find suitable site

- solar energy – solar panels 'catch' and store the energy of the sun.

We all know about nuclear energy:

- fission energy, where the nuclei of items break into smaller nuclei and

- (greater) power of fusion energy, where the nuclei bind together into larger ones.

Research is underway to develop technologies like such as hydrogen-powered cars and solar-powered generators.

The conservation of energy

KEY POINT

Energy can neither be created nor destroyed but can only be transformed from one form to another.

Hence, the total amount of energy in the universe always remains constant or the total energy of a closed or isolated system always remains constant.

For example, if a bullet hits a target and enters into it upto some distance, then a part of its K.E. is converted into heat energy. The remaining part may give K.E. to the target, i.e., if a suspended mass is hit by a bullet then some of the K.E. of the bullet is converted into heat and the remaining kinetic energy will cause the suspended mass and embedded bullet to swing upwards, gaining potential energy. The mass will then begin to oscillate.

Summary

Work : Work done is the product of force and the distance through which a body moves in the direction of force.

Unit : **Joules**

Work is a scalar quantity.

Power : Rate of doing work is called power.

Unit : **Watts**

Energy : Energy is defined as the capacity of doing work.

Kinetic Energy : Kinetic energy is a scalar property, which is associated with the state of motion of a moving body. The kinetic energy of a body is always positive.

Potential Energy : The energy stored in a body or a system by virtue of its position or configuration in a field is called potential energy. A moving body may or may not have potential energy.

Various types of potential energy are :

● Elastic potential energy, e.g.: a compressed spring, a stretched bow.

● Gravitational potential energy, e.g.: a planet moving around the sun, water stored in water tanks.

The potential energy stored in a spring : In stretching or compressing a spring, some work is done on it. This work is stored in it in the form of potential energy.

If 'x' is the compression or extension in the spring by applying a force F, then $F = -kx$. (where k is called the **spring constant**, or force constant of the spring)

KEY POINT

If a spring is stretched from $x = x_1$ to $x = x_2$, then the amount of work done, $W = \frac{1}{2} k (x_2^2 - x_1^2)$

This is stored in the spring in the form of potential energy.

Sample IGCSE questions

1. A train accelerates from rest (0 m/s) to its maximum speed of 60 m/s in a time of 120 s.

(a) Calculate the acceleration of the train. **[3]**

> *Acceleration = change in velocity ÷ time taken* ✓
> *= 60 m/s ÷ 120 s* ✓
> *= 0.5 m/s² ✓*

> *Take care with the units of acceleration. A common error at IGCSE is to give the unit as m/s instead of m/s².*

(b) The mass of the train is 350 000 kg.

(i) Calculate the size of the force needed to accelerate the train. **[3]**

> *force = mass × acceleration* ✓
> *= 350 000 kg × 0.5 m/s²* ✓
> *= 175 000 N* ✓

> *There is one mark here for recall of the formula; the second mark is for identifying the quantities correctly and the final mark is for obtaining the correct answer and giving the correct unit.*

(ii) Explain why the acceleration of the train decreases as its speed increases. **[3]**

> *Resistive forces act on the train ✓.*
> *These forces increase as the train's speed increases ✓.*
> *Reducing the size of the unbalanced force acting on the train ✓.*

> *The size of the unbalanced force is the difference between that of the driving force and that of the resistive forces, since these act in opposite directions.*

(c) A jet aircraft of the same mass as the train accelerates from rest to a speed of 60 m/s before it takes off from the ground.

(i) Explain why the aircraft needs a much greater force than the train to reach the same speed. **[3]**

> *The aircraft needs to reach take-off speed in a shorter distance ✓.*
> *So it needs a greater acceleration ✓.*
> *A bigger force is needed to cause the greater acceleration ✓.*

> *You are not expected to know anything about aircraft to answer this question. You are being asked to use your understanding of f = m × a to evaluate the information that you have been given.*

(ii) Suggest why it is important that aircraft do not take off with far more fuel than is needed for the journey. **[1]**

> *Excessive mass increases the distance required for take-off OR increases the fuel consumption on the journey ✓.*

Sample IGCSE questions

2. In a game of tennis, a ball leaves a racket travelling horizontally at a speed of 34 m/s from a height of 2.8 m above the ground.
The diagram shows the path of the ball after it is hit.

34 m/s

2.8 m

The effect of air resistance on the ball can be neglected.

(a) What happens to the **horizontal velocity** and **vertical velocity** of the ball after it leaves the racket? **[2]**

The horizontal velocity does not change ✓ and the vertical velocity increases ✓.

> The fact that the horizontal velocity does not change is inferred from the statement that air resistance can be neglected.

(b) Explain the shape of the path followed by the ball. **[3]**

The ball travels equal distances in equal time intervals horizontally ✓ but increasing distances vertically as its speed increases ✓ causing the path to become steeper ✓.

(c) Calculate the time it takes for the ball to reach the ground.

The value of free-fall acceleration, g = 10 m/s². **[3]**

$s = ut + \frac{1}{2}at^2$ *with* $u = 0$ ✓
so $t = \sqrt{(2s/a)}$ ✓
$t = \sqrt{(2 \times 2.8\ m/10\ m/s^2)} = 0.75\ s$ ✓

> It is important to realise here that **vertically** the initial velocity is zero. A common error is to use data about the horizontal motion to answer a question about the vertical motion.

(d) How far does the ball travel horizontally in this time? **[3]**

distance travelled = speed × time ✓
$= 34\ m/s \times 0.75\ s$ ✓
$= 25\ m$ ✓

> The time is the only variable that has the same value for both the horizontal and the vertical motion.

(e) Calculate the **vertical velocity** of the ball as it reaches the ground. **[3]**

$v = u + at$ *with* $u = 0$ ✓
$= 10\ m/s^2 \times 0.75\ s$ ✓
$= 7.5\ m/s$ ✓

> Remember, the correct unit is usually required to gain the final mark in a calculation.

Exam practice questions

1. (a) The diagram shows the forces between the wheel of a bicycle and the road when the bicycle is accelerating forward.

(i) Force A is the backward push of the wheel on the road.

Write a similar description of force B. [2]

(ii) Which option describes the relative sizes of these forces correctly?

A Force A is greater than force B.

B The forces are equal in size.

C Force B is greater than force A.

Write down the letter of your choice. [1]

(iii) Explain why wet leaves between the wheel and the ground may cause the wheel to spin round. [2]

(b) The mass of the cycle and cyclist is 90 kg.

The forward force has a value of 120 N.

The resistive force has a value of 60 N.

(i) What is the main resistive force that acts on the cyclist? [1]

(ii) Calculate the size of the unbalanced force on the cyclist and state its direction. [2]

(iii) Calculate the acceleration of the cyclist. [3]

(iv) Explain how the acceleration of the cyclist changes as the speed of the cycle increases. [3]

2. The graphs show how the thinking distance and braking distance of a car are related to its speed.

Exam practice questions

(a) Describe the relationship between:

    (i)   thinking distance and speed   **[1]**

    (ii)   braking distance and speed.   **[1]**

(b) Use the graphs to give the values of the:

    (i)   thinking distance at a speed of 26 m/s.

    (ii)   braking distance at a speed of 26 m/s.

    (iii) stopping distance at a speed of 26 m/s.   **[3]**

(c) Write down TWO factors that could affect:

    (i)   the thinking distance

    (ii)   the braking distance of a vehicle.   **[2]**

3. The graph shows how the velocity of a ball changes after it is thrown vertically upwards.

(a)   **(i)**   Calculate the acceleration of the ball.   **[3]**

    (ii)   What is the direction of the acceleration?

          Explain how you can tell.   **[2]**

    (iii) The mass of the ball is 0.020 kg.

          Calculate the size of the force required to cause this acceleration.   **[3]**

(b)   **(i)**   After what time shown on the graph did the ball change direction?

          Explain how you can tell.   **[2]**

    (ii)   Use the graph to work out the height that the ball reached.   **[3]**

4. A Saturn V Moon rocket burns fuel at the rate of 13 000 kg each second. The exhaust gases have a speed of 2500 m/s.

(a) Calculate the change in momentum of the exhaust gases in 1 s.   **[3]**

(b) The rocket has a mass of 2.7×10^6 kg. Calculate the change in its speed per second.   **[3]**

(c) It continues to burn fuel at the same rate. Explain why its acceleration increases as more fuel is burned.   **[2]**

Exam practice questions

5. The diagram shows two boys on roller blades facing each other.

30 kg   50 kg

The smaller boy pushes the larger boy with a force of 150 N. The larger boy then moves off at a speed of 1.5 m/s.

(a) What is the size of the force on the smaller boy? State its direction. [2]

(b) Explain, in terms of momentum, why the smaller boy moves backwards. [3]

(c) Calculate the speed of the smaller boy. [3]

6. A ball is travelling through the air.
Its horizontal speed is 15 m/s.
Its vertical speed is 10 m/s.

Use a vector diagram to work out the speed of the ball through the air, and the direction of travel compared to a vertical line. [3]

7. A dart is thrown horizontally at a dartboard.

1.8 m

The horizontal speed of the dart is 12.4 m/s and the distance from the point of release to the dartboard is 1.8 m.

(a) Describe and explain the path of the dart as it travels through the air. [3]

(b) Calculate the time it takes for the dart to reach the dartboard. [3]

(c) Calculate the vertical distance travelled by the dart as it travels through the air.

The value of free-fall acceleration, $g = 10$ m/s². [3]

Exam practice questions

8. The diagram shows the circular orbits of two satellites.

(a) Explain why an **unbalanced force** is required to keep a satellite in a circular orbit.   [2]

(b) Describe the force that keeps a satellite in orbit.   [2]

(c) What three factors determine the size of the unbalanced force required?   [3]

(d) The orbital time of a satellite in a polar orbit is much less than that of a satellite in a geostationary orbit. Suggest two reasons for the same.   [2]

(e) Here is some data about a surveillance satellite in a circular orbit.

radius of orbit = 7.1×10^6 m

speed of satellite = 7.5×10^3 m/s

mass of satellite = 1.2×10^3 kg

   (i) Calculate the orbital time of the satellite.   [3]

   (ii) Calculate the centripetal acceleration of the satellite.   [3]

   (iii) Calculate the value of the unbalanced force that acts on the satellite.   [3]

   (iv) What is the direction of this force?   [1]

9. A golf ball has a mass of 0.046 kg.
    After being struck by a golf club, it leaves the tee at a speed of 60 m/s.

(a) Calculate the change in momentum of the golf ball.   [3]

(b) The golf club is in contact with the ball for a time of 4.0×10^{-4} s.
    Calculate the force exerted on the ball.   [3]

(c) After hitting the ball, the club carries on moving with no noticeable change in speed. Suggest the reason for this occurence.   [2]

Thermal physics

The following topics are covered in this section:

- **States of matter**
- **Thermal expansion**
- **Energy transfer and insulation**
- **More about gases, pressure in fluids**
- **Temperature and thermal capacity**

2.1 States of matter : the molecular view

LEARNING SUMMARY

After studying this section you should be able to:

- **distinguish between solids, liquids and gases on the basis of physical properties**
- **understand the concepts of melting point and boiling point**
- **enumerate the factors affecting rate of evaporation**
- **describe brownian motion in terms of random molecular bombardment**
- **explain diffusion through everyday examples.**

Melting and boiling points

Matter exists in three familiar phases or forms—solids, liquids and gases. There is a fourth form, known as **plasma** (which is really an ionised gas), but this occurs mostly in extreme conditions and we do not need to focus on it here.

Solids are hard to the touch and they have a *fixed shape*. To change their shape, we need to exert a good deal of energy on them by beating them or heating them.

On a microscopic level, the molecules of the solid cannot move around each other, as they are held in their respective positions by strong bonds. Hence, they can only vibrate about their positions.

As we add energy to the solid by heating it, the vibrational energy of the molecules increases. When they reach a certain temperature (the **melting point**), they have enough energy to move around. The intermolecular bonds are still strong and the molecules are not independent, hence they cannot move away from each other. They can only 'slide' around each other. This is why a liquid can flow and take up the shape of the container into which it is poured.

If we continue to heat the liquid, its molecules acquire more and more energy. This is why a puddle of rainwater dries up on a sunny day. The sun's heat warms up the water and when a molecule acquires enough energy, it leaves the surface of the liquid as a vapour. The process is called **evaporation**.

Evaporation takes place over a wide range of temperatures, but occurs at a faster rate as the temperature increases. This is due to the fact that individual molecules have a greater probability of acquiring enough energy to leave the surface of the liquid (the other molecules near the surface try to hold them back due to the intermolecular forces) at higher temperature. Actually, the rate of evaporation increases when any of these factors increase:

- temperature
- surface area of liquid
- draught, or wind passing over the surface.

So, a washerman will tell you that the quickest way of drying clothes is to hang them out on a sunny day when there is a strong breeze, preferably without folding them so that the whole of their area is exposed.

Evaporation has a cooling effect. Clearly, if an energetic molecule escapes from the surface of a liquid, it carries its energy away with it. You must have noticed that when you sweat, the drops of perspiration evaporate from your skin, leaving it so cool that you may even shiver sometimes.

If we continue to heat a liquid, a certain temperature (the **boiling point**) is reached when the intermolecular bonds are broken and the molecules can move independently of each other – the liquid has become a gas. Unlike the solid and liquid states, where the molecules are close together and therefore incompressible, the gaseous state is characterised by the rapid, random, straight line motion of the molecules. A gas expands to fill its container because the molecules have a lot of energy, they are free to move around the whole of the container, bouncing off the walls and (less often) each other. This explains pressure on a molecular scale – a car tyre remains inflated because the molecules of air inside it are bombarding against its interior wall. Similarly, a piston in an engine is pushed violently upwards due to the bombardment of many high energy gas molecules.

Gases can be compressed. A bicycle pump compresses air and pushes it into the tyre, which becomes harder as the air is compressed and the internal pressure increases.

Brownian motion and diffusion

The rapid, random, straight line motion of gas molecules is known as **brownian motion**. It can be seen when smoke particles are viewed under a microscope. The smoke particles move in a zig-zag manner, as the energetic molecules of air collide with them haphazardly.

If a small quantity of perfume is sprayed at the entrance of a closed room, the smell will permeate throughout the room, taking a minute or so to reach the furthest corner of the room. This process is known as **diffusion** and occurs due to the collision of perfume molecules with air molecules. The zig-zag brownian motion eventually causes the perfume to spread out throughout the container (in this case, the room).

Liquids and gases are collectively known as fluids, as they both *flow*.

The properties of solids, liquids and gases have been summarised in **Table 2.1** for quick reference.

SOLIDS	LIQUIDS	GASES
incompressible	incompressible	can be compressed
fixed shape	takes on shape of the container; has a surface	fills its container
fixed volume	fixed volume	no fixed volume
static	can flow	can flow

2.2 More about gases, pressure in fluids

LEARNING SUMMARY

After studying this section you should be able to:

- *define the properties of an ideal gas*
- *understand the relation between the pressure applied and volume of a gas: Boyle's law*
- *convert temperatures from Celsius to Kelvin scale and vice versa*
- *understand the working of barometers and manometers*
- *express pressure as a function of depth and density of the liquid.*

Pressure and volume

A common error at IGCSE is to state that gas pressure is due to the collisions between the particles.

Gases consist of large numbers of particles in constant motion. The motion of an individual particle is random, it is constantly changing in both speed and direction as it collides with objects and other particles. Unlike the pressure due to a solid object, gases exert pressure equally in all directions.

KEY POINT

Gas pressure is a result of the forces exerted when the particles collide with the walls of the container.

Generally, when we talk about gases, we usually mean perfect or **ideal** gases, where the molecules are assumed:

- to occupy no space — this is almost true, as the amount of space taken up

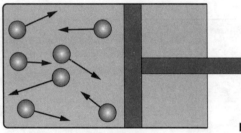

by the molecules is negligible compared to the distance between them

● to be point objects. Actual molecules are not point objects. Methane (CH_4), for example, is not a nice point molecule – it has a carbon atom with four hydrogen atoms bonded to it in the shape of a tetrahedron.

Having said that, all gases behave almost ideally over a wide range of temperatures.

We have said that gases fill up the container in which they are enclosed. If a quantity of gas is placed in a piston (which is a cylinder with a tight but smooth fitting moveable top) and the volume is slowly reduced, as in **Fig. 2.1**. Clearly, the amount of collisions suffered by the swiftly moving molecules, per second, will increase. Hence if the volume is decreased, we register an increase in pressure.

> **Remember that the temperature must be constant for Boyle's law to hold good!**

Fig. 2.1

Acually, for an ideal gas, **Boyle's law** holds:

> **KEY POINT**
>
> **For a constant temperature, pressure x volume = constant,**
>
> **or pV = constant.**

> **Make sure that if pressure is given in N/m², the volume should also be given in m², otherwise you will have to convert the units so that they are compatible.**

>
> **EMAMPLE**

At constant room temperature, the pressure of a gas is 0.25 N m^{-2}, when it occupies a volume of 100 cm³. If the gas is now compressed so that its volume becomes 40 cm³, what will its new pressure be?

Ans: Since the temperature is constant, pressure x volume will be constant.

Hence, $p_1 V_1 = p_2 V_2$, so $0.25 \times 100 = p_2 \times 40$.

New pressure = 25 ÷ 40 = 0.625 N/m².

> **Remember: Pressure is defined as force per unit area, so it is measured in N/m² or pascals (Pa).**

Temperature may be given in degrees celsius, °C, (another name for centigrade) or on the absolute temperature scale in Kelvin, K. Note that there is no ° symbol in the Kelvin unit.

> **Real gases would liquify before reaching absolute zero temperature. The absolute scale is based on an idealised gas model.**

0 K (Zero Kelvin) is as cold as you can go – at this temperature, the molecules of a solid cease to vibrate and so there is no energy remaining in the substance. It is equivalent to a temperature of –273°C.

To convert from celsius to kelvin, add 273 degrees. Hence, a temperature of 100 °C = 373 K.

To convert from kelvin to celsius, subtract 273. So a temperature of 800K equals a temperature of 527 °C (800–273).

The diagram, **Fig 2.2**, shows the relationship between these temperature scales.

-273	temperature / °C	0	100
absolute zero		ice point	steam point
0		273	373

temperature /K **Fig. 2.2**

> Doubling the temperature of a gas doubles the mean kinetic energy of the particles, but this does not mean that speed is doubled.

When temperature is measured in kelvins:

● the mean kinetic energy of gas particles is proportional to their temperature

● the pressure of a gas is proportional to its temperature.

The last point means that when the kelvin temperature of a gas is doubled, its pressure also doubles.

PROGRESS CHECK

1. In a machine working at a constant temperature of 400K, a piston enclosing 360 cm³ of air at a pressure of 2 Pa compresses the air to a volume of 80 cm³.

(a) Write the temperature 400K in degrees celsius.
(b) What is the pressure of the gas after compression?

2. Explain how evaporation occurs from the surface of a liquid.
3. Draw the graph between PV and P at constant temperature for the given mass of a gas.

3.

AB is the required straight line graph. PV is a constant, as long as the temperature is constant. Here, as the pressure increases, volume decreases. So PV is a constant for all pressures.

2. A liquid molecule must have certain minimum kinetic energy or velocity to overcome the surface attractive forces, in order to escape from the liquid surface. The average velocity of most of the molecules is less than this limiting velocity. However, some molecules can have this high velocity. If such molecules happen to be on the surface, they can easily escape the surface. This is how evaporation occurs.

1. (a) 127 °C; (b) 9 Pa;

Pressure and temperature

Experiments show that increasing the temperature of a gas results in an increase in the pressure. This is because:

● increasing the temperature increases the average speed of the particles
● this results in the collisions becoming more frequent
● the average force exerted during the collisions is greater.

When the pressure of a gas is plotted against its temperature, the result is a straight line shown in **Fig. 2.3**. The straight line does not go through the origin because the Celsius temperature scale is not an absolute scale. On an absolute scale of measurement:

- zero is the minimum possible value
- 20 units have twice the value of 10 units.

This graph can however be used to locate the absolute zero of temperature and to set up a new scale that does measure temperature in absolute terms.

Fig. 2.3

At the absolute zero of temperature the movement of the particles of a gas is at a minimum, and so is the pressure they exert. The diagram, **Fig. 2.4**, shows what happens when the graph is extrapolated to find the temperature at which a gas exerts minimum pressure.

Fig. 2.4

> **KEY POINT**
>
> The relationship between the pressure of a gas and its kelvin temperature is:
>
> $$\frac{P_1}{T_1} = \frac{P_2}{T_2}$$
>
> This relationship can also be written as $P \times T$ = constant.

The gas equation

At a constant temperature, the pressure of a gas is inversely proportional to its volume. The equation that describes this ($P_1 V_1 = P_2 V_2$) can be combined with the equation above to give a single equation. This equation applies to a situation where there are changes in the pressure, volume and temperature.

> **KEY POINT**
>
> The gas equation is:
>
> $$\frac{P_1 V_1}{T_1} = \frac{P_2 V_2}{T_2}$$

Measuring pressure

> **KEY POINT**
>
> **Barometers** are used to measure atmospheric pressure. **Manometers** are used to measure pressure difference.

Barometers are of two types:

Liquid (mercury) barometer

In 1643, *Torricelli* first measured atmospheric pressure by investigating the height of a mercury column that could be supported, with the atmosphere exerting a downward pressure on the surface of mercury filled in a trough, **Fig. 2.5**.

Fig. 2.5

Aneroid barometer

Here, air pressure causes a spring to uncoil. The spring carries a pointer which moves across a dial, that is calibrated by the manufacturers to read atmospheric pressure.

Fig. 2.6

Manometer

A simple manometer is a U-tube that contains water or a denser liquid, depending on the pressure to be measured. One side of the manometer is connected to a gas source and the gas pressure forces the liquid down on that side of the tube. Consequently, the liquid on the other side rises. The gas pressure can be found from the difference in these levels.

Fig. 2.7

Pressure in a liquid

In a liquid, pressure increases with depth according to the formula:

pressure = density x gravity x depth = ρgh

It is the difference in pressures between the lower and upper surfaces of an object, when immersed in a liquid, that causes the **upthrust** or upwards force on a liquid. Since the pressure in a liquid increases with depth, large river dams are not built as a vertical concrete wall as they would never withstand the pressure. Their cross-section is in the shape of a curving wall that gets progressively thicker from top to bottom. The curve is a stronger structure than a horizontal barrier, just as an archway is stronger than a horizontal beam.

At any given depth, the pressure exerted by a liquid acts equally in all directions. This fact is exploited best by the brake systems of vehicles. When a driver presses on the brake pedal of a car, a non-corrosive liquid is pushed outwards to a cylinder in each wheel where a piston moves to press the brake discs against the wheels. Since the pressure is transmitted via the liquid, it is the same in every direction, i.e. at every wheel. Hence, the car will not pull to one side as it would have if the force at each wheel varied.

The variation of pressure with height can be observed by carrying out a very simple experiment. If you take a tall can and fill it with water, and then punch holes in it, you will see that the water comes out at different rate. The stream of water at the bottom emerges with greater force than that at the top, due to the greater pressure at the bottom.

Fig. 2.8

If you take another can with some water in it and either boil the water or allow the water to run away while immersed in a sink, you will see that the can will collapse. This is because there is vaccum inside the can and the external air pressure pushes the walls of the can inwards.

---

1. Neglecting atmospheric pressure, find the pressure on a diver who is 12 m below the surface of a lake.

2. A knife of mass 1 kg, that is made of steel, will sink if thrown into the sea, but an oil-tanker ship, loaded with a tonnage of 40 Mkg (that's 40 million kilograms or 40 thousand tonnes) floats. Explain.

3. A man is sitting in a small boat in the middle of a swimming pool. A large stone is placed in the boat. If he throws the stone overboard, will the water level in the pool rise or fall?

PROGRESS CHECK

1. Density of water = 1,000 kg/m³; g = 9.8 m/s²; depth = 12 m.
   Hence, pressure = 1,000 × 9.8 × 12 = 117.6 kPa
   He would not survive a pressure much greater than this, without a protective steel diving suit.

2. The ship's mass is spread out over an extremely large volume as the ship is, in effect, a hollow shell and so its density per unit volume is much lower than that of the solid knife. It is density, and not mass alone, that determines whether the object sinks or floats.

3. When the stone is in the boat, it displaces its own weight of water, W. When it is thrown overboard, it sinks and thus displaces its own volume of water, V. However, V is less than W and so the water level falls.

2.3 Thermal expansion

LEARNING SUMMARY

After studying this section you should be able to:

- ***describe the expansion of solids, liquids and gases qualitatively***
- ***delineate the role of thermal expansion in everyday examples***
- ***quote some applications of thermal expansion.***

- When a solid is heated, it expands.

- When a liquid is heated, it turns into a gas.

- When a gas is heated, it exerts greater pressure on the walls of its container.

Here, we look at the thermal expansion of solids. We shall start with some interesting facts about expansion.

The fuselage of high-performance supersonic planes can increase in length by more than ten centimetres. As the plane pushes through the air, it does work against air resistance. This causes the fuselage to heat up and expand.

In Australia, the Sydney Harbour Bridge regularly expands on hot days and contracts in the colder nights. The bridge is mounted on a system of rollers that allow for its change of length. Whenever you pass under a large bridge or flyover, look at its structure and see how allowance has been made for expansion.

In tropical countries, railway lines have been known to buckle and deform as the steel tries to stretch but is restrained by the bolts which fasten it to the sleepers, or ties. Next time you look at a railway track, notice that it is not a continuous piece but consists of many lengths of steel that are separated from each other by gaps, which allow for expansion in hot weather.

On a molecular scale, when a solid heats up, its constituent particles (usually molecules, but atoms in the case of an element such as iron) have a greater average energy. They vibrate more energetically and the intermolecular spacing increases by a small amount, depending on the substance. On a large scale, this is quite observable.

It is for this reason that when a steel wheel is mounted on an axle, the wheel is heated (and so the central hole enlarged) and the axle chilled, so that its cross-sectional area is reduced. The two are then fitted together quite easily. When the system reaches room temperature, the wheel has contracted and the axle has expanded, making the seperation of the two impossible.

Likewise, if the aluminium lid of a jar is too tight to open, it is amazing the difference that holding the lid in hot water will make (hold the glass jar in a towel, just in case thermal forces are set up in the glass that cause it to break).

Some electrical circuits make good use of a bimetallic strip **Fig. 2.9** where two metals, that expand at different rates, are welded together. Upon heating, one metal lengthens more than the other which tries to restrain it. The bimetallic strip thus bends. It can be used to make and break a circuit intermittently, acting like a switch. The bimetallic strip was often used in the indicator circuits of cars, allowing them to flash on and off periodically. It is also used in thermostat controls, to maintain a constant temperature and in fire alarms, where increasing heat causes the strip to bend more and more until at a given temperature, it touches a circuit and allows a current to flow through it and ring a bell.

expands and contracts quickly

expands and contracts slowly   **Fig. 2.9**

2.4 Temperature and thermal capacity

LEARNING SUMMARY

After studying this section you should be able to:

- *identify tempreature as a physical property of a system*
- *understand the concept and need for fixed points of a temperature scale*
- *describe the structure and action of liquid-in-glass thermometers*
- *distinguish between mercury and alcohol as potential candidates for use in thermometers*
- *understand the concept of heat capacity and use it to relate the heat exchange with change in temperature of a body*
- *elucidate the molecular view of latent heat.*

Temperature

It must be remembered that heat always flows from a body at higher temperature to a body at lower temperature.

The degree of hotness or coldness of a body is measured by a quantity, which is called temperature.

It may also be defined as the property of a body which governs the flow of heat.

Heat

Energy that flows from one body to another, due to temperature difference, is defined as heat.

The heat contained in a body depends upon its temperature and mass, as well as the material from which it is made.

Thermometer

A device which can measure the degree of hotness or coldness (i.e., temperature) of a body accurately and reliably, is known as a thermometer.

Thermal equilibrium

If two bodies are at the same temperature, then there will be no transfer of heat between them. The bodies are said to be in thermal equilibrium. In such a case, there is no temperature difference between any two parts of the body or the system.

Thermometric substance

The substance whose property is utilised for measuring temperature is called a thermometric substance and the property used to measure temperature is called thermometric property.

Type of thermometers

Thermometers are categorised as:

- liquid–in–glass thermometer
- constant volume gas thermometer
- constant pressure gas thermometer
- thermocouple
- pyrometer.

There are many more, but here we will study only the liquid–in–glass thermometer.

The principle of a liquid-in-glass thermometer

Liquids expand on heating and their expansion is generally uniform and regular. This property of uniform expansion of liquids is used as a thermometric property.

Lower fixed point and upper fixed point

There are two fixed temperatures on every thermometer and these points are called fixed points.

Lower fixed point

This is the temperature at which pure ice melts, under standard atmospheric pressure.

Upper fixed point

This is the temperature at which pure water boils, under standard atmospheric pressure.

Water is used because it is freely available. Hence, one can calibrate (or accurately mark the scale) the thermometer by immersing the liquid bulb at the bottom of the thermometer in ice (to get the lower fixed point) and then in steam (to get the upper fixed point).

> The Celsius scale is a relative scale, temperatures are measured relative to the freezing point of water.

The Celsius scale of temperature

On the celsius scale, the lower fixed point is 0°C and the upper fixed point is 100°C. An older scale (now falling out of use) is the Fahrenheit scale, which has its fixed points based on the melting and freezing of sea water rather than pure water.

Fig. 2.10 Thermometers

(a) Thermometer used in science experiments

(b) Clinical thermometer

(c) Wall thermometer

The usual liquids used in a thermometer are either mercury or alcohol. The internal space in the thermometer tube is very thin – it is a capillary tube, and so quite sensitive, easy to handle and inexpensive.

Mercury is extensively used because:

- it is the only metal that is a liquid over a wide range of temperatures above and below room temperature

- it is non-corrosive

- it has a reasonably high coefficient of expansivity, i.e. it expands by a measurable amount for a small rise in temperature.

Alcohol, which is colourless, is often used in a thermometer because it is cheaper than mercury. A red or blue colouring is added so that the alcohol thread is clearly visible. Alcohol does not have as high a coefficient of expansivity as mercury, nor is it as responsive. Being a metal, mercury responds very quickly to heat changes.

A doctor uses a medical thermometer, which operates over a small range of temperatures around the normal temperature of the human body. Just above the bulb containing the liquid, there is a constriction so that when it is removed from the patient's mouth, the thread is trapped and cannot shrink downwards towards the colder room temperature.

Specific heat capacity

All objects have **internal energy** due to the movement of the particles. In solids and liquids there is a constant interchange between kinetic and potential energy as the particles vibrate and jostle against each other. In gases the internal energy is kinetic, and potential. The potential energy the particles in a gas have is gained when they change from a liquid to a gas.

> Temperature is a measure of the internal energy of the particles that make up an object.

Changing the temperature of an object involves changing its internal energy. This can be done by heating or cooling. The amount of energy depends on:

- the material the object is made from
- the mass of the object
- the temperature change.

> The term specific is used to refer to 1 kg of material.

These factors are all taken into account in the concept of **specific heat capacity**:

- specific heat capacity, c, is a property of a material
- it is the amount of energy required to change the temperature of 1 kg of the material by 1°C or 1 K.
- the units of specific heat capacity are J/kg °C or J/kg K.

> Although the values of temperatures differ by 273 when measured on the Celsius and kelvin scales, a temperature difference has the same value no matter which scale is used.

> **KEY POINT**
>
> The energy transfer when an object is heated or cooled is calculated using the relationship:
> energy transfer = mass × specific heat capacity × temperature change
> $$E = m \times c \times \Delta\theta$$
> where $\Delta\theta$ is the temperature change in °C or K.

Thermal capacity

The thermal capacity of a body is defined as the amount of heat required to raise the temperature of the whole body by 1K.

Two bodies of same material have same specific heat capacity but if their masses are different, then they will have different thermal capacities. For example, a cup of hot water that is near boiling point has the same specific heat capacity as a bath of warm water at a lower temperature (they are both composed of water). However, the bath has greater thermal capacity, i.e. it contains greater thermal energy due to its much greater mass.

Actually water has a very high specific heat capacity, so it takes a lot of energy to raise its temperature. For instance, boiling a kettle of water requires a god deal of energy. It has a specific heat capacity of 4200 J/ kg °C. This means that it takes 4.2 kJ of heat energy to raise the temperature of 1 kg of water through 1°C, (although in practice the value will be higher because there will always be heat losses to the surroundings also).

Example: A platinum ball of specific heat capacity 168 J kg °C is heated in an insulated vessel from 25°C to 95°C. If the mass of the ball is 10 g, calculate the heat energy supplied to the ball.

Ans: Energy required = mass x temperature rise x specific heat capacity, i.e. 0.01 x (95 – 25) x 168 = **117.6 J**, (or 120 J to 3 significant figures).

Melting and boiling

If we take a block of ice at -10 °C and heat it, its temperature will begin to rise. The heat energy we are supplying is stored in the block. Due to this energy, the molecules vibrate more quickly and this is registered as a temperature rise. But when we reach the freezing point of water (also know as the melting point of ice), at 0 °C, the temperature ceases to rise, even though we keep heating the ice-block. Where, then, is the heat going?

As the solid changes into a liquid, the molecules are able to do more than plain vibration – they can move *around each other*. Even though strong bonds hold them together, there is a fundamental difference in molecular motion. This is why the non-flowing ice changes to water, which has the ability to flow. The heat we are adding is seen as a **change of state** rather than a change of temperature. Since we do not see a temperature rise, the heat we supply is called a 'hidden heat' or **latent heat**.

Similarly, at 100 °C, the temperature remains at the boiling point of water until all the water has changed into steam. Only then will the temperature begin to climb again, as the steam gets hotter.

KEY POINT
There are two latent heats, then:
• when the solid changes to a liquid (or vice-versa), we have the latent heat of fusion
• when the liquid changes to a gas (or vice-versa), we have the latent heat of evaporation.

Latent heat of fusion

This is the quantity of heat required to convert 1 kg of a solid at its melting point into liquid, without any change of temperature.

ab – latent heat of fusion

 (state = solid+liquid)

cd – latent heat of vapourisation

 (state = liquid+vapour)

bc – liquid state

Fig. 2.11

Latent heat of vaporisation

It is the quantity of heat required to convert 1 kg of a liquid at its boiling point into its vapour, without any change of temperature.

Latent heats of vaporisation are higher than latent heats of fusion as complete bond-breaking occurs when the state changes from liquid to gas.

The latent heat of fusion can be found using a known mass of crushed ice in a calorimeter (an insulated copper can) which contains an electric heater. Then, the ice is heated until it has changed into water, and a record of time is kept using a stopwatch. When all the ice has melted and the temperature starts to rise, the power of the heater and the time of heating is noted. The product of these gives the energy supplied to the ice. By knowing the mass of ice, we can work out the latent heat of fusion.

 To find the latent heat of evaporation, a similar set-up is used, except that here we bring a known mass of water to the boil and record the time for phase change, stopping our watch when all the water has evaporated.

The units of latent heat are J/kg. Temperature does not come into account, as there is no temperature rise during the phase change.

PROGRESS CHECK

1. State two reasons why the pressure of a gas increases when its temperature rises.
2. At what temperature do the particles of a gas have twice as much kinetic energy as at 20°C?
3. What kind of energy do particles gain when they change from a liquid to a gas without a change in temperature?

1. The collisions are more frequent. The average force during collisions is greater because the average speed of the particles is greater; 2. 313°C or 586 K; 3. The particles gain potential energy.

2.5 Energy transfer and insulation

LEARNING SUMMARY

After studying this section you should be able to:

● *explain the mechanisms of energy transfer by conduction, convection and radiation*
● *understand the role of trapped air in insulating houses and people.*

The nature of the surface

All objects emit and absorb energy in the form of **infra-red radiation**, electromagnetic waves with wavelengths longer than light but shorter than microwaves. The rate at which energy is emitted depends on the temperature and the nature and colour of the surface. The rate at which energy is absorbed depends only on the nature and colour of the surface.

> The range of wavelengths emitted depends on the temperature of the object. Hot objects also emit light.

KEY POINT
> Dark, dull surfaces are good emitters and absorbers of infra-red radiation. Light, shiny surfaces are poor emitters and absorbers of infra-red radiation.

Infra-red radiation is reflected in the same way as light. For this reason:

● marathon runners are often given aluminium foil capes to wear after a race; *crucial to conserve heat* it reflects back the infra-red radiation emitted by their bodies

● even though it conducts heat, aluminium foil keeps food taken out of an oven hot, as it reduces energy loss by radiation

● in hot countries, houses are often painted white to reduce the energy absorbed from the Sun's radiation.

Examples of objects that are painted black to maximise the absorption or emission of radiant energy include:

> Solar panels used to generate electricity are also painted black to maximise the energy absorbed from the Sun's radiation.

● the pipes at the rear of a refrigerator or freezer, as the coolant flowing through these pipes is warm and needs to lose the energy absorbed from the inside of the cabinet

● solar heating panels that absorb radiant energy from the Sun and use it to heat water.

Moving fluids

The movement of a gas or a liquid in a **convection current** is due to parts of the fluid having different **densities**. When a liquid or a gas is heated:

● the particles gain more kinetic energy, causing the fluid to expand

● this results in a decrease in the density of the fluid

● the warmer, less dense fluid rises and is replaced by colder, denser fluid.

> The density of an object or a fluid is the mass per unit volume; it is a measure of how close-packed the particles are.

The reverse happens when a fluid is cooled. Central heating radiators rely on upward-driven convection currents to circulate the warm air in a room while refrigerators rely on downward-driven convection currents to keep the contents cool. These are shown in the diagram, **Fig. 2.12**.

Fig. 2.12

Conduction by particles

> Particles interact with their neighbours through their vibrations; they exert repulsive and attractive forces as they move together and then apart.

All materials allow energy to pass through them by **conduction**. As particles become more energetic, some of this energy is transferred to neighbouring particles as they interact. The particles of a gas are more widespread than those of a solid or a liquid, so interactions between them are less frequent. This is why gases are poor thermal conductors.

Metals are better conductors than non-metals because the free electrons responsible for conduction of electricity in metals also play a role in thermal conduction. These electrons:

● move randomly at high speeds

● travel relatively large distances between collisions with the metal ions

● transfer energy rapidly from hot areas of the metal when they move to cooler areas by diffusion.

Insulating buildings and bodies

Of the three methods of energy transfer described above, the greatest energy loss from warm buildings and bodies occurs through conduction and convection. Energy loss by radiation is more significant when an object is considerably warmer than its surroundings.

Energy is transferred through a glass window, a brick wall and through clothing by conduction. In modern houses the external walls consist of a brick outer wall and a breeze block inner wall. An air gap, called a cavity, separates the two. Energy passes from the warm inside to the cold outside by:

> Energy loss by radiation from the head can be significant. Hair is a good insulator, so people who are bald should always wear a hat in cold weather to reduce this energy loss.

● conduction through the inner wall

● convection through the cavity

● conduction through the outer wall.

Energy flow through the wall is reduced by cavity wall insulation. This stops the convection in the cavity by trapping pockets of air. Energy can now only flow through the cavity by conduction. Since air, like all gases, is a poor conductor, the energy flow is much less. Energy flow through the walls of a house is shown in the diagram, **Fig. 2.13**.

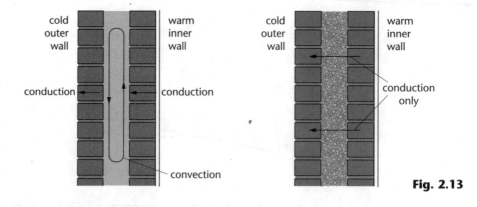

Fig. 2.13

Energy flows through a double-glazed window in a similar way. As it is impractical to use another material in the gap between the panes, these should be placed close together. A narrow gap does not allow enough room for convection currents to flow between the glass panes.

Loft insulation is the most cost-effective way of insulating a house. Energy flows through an uninsulated loft in a similar way to an uninsulated external wall:

- energy is conducted through the plasterboard ceiling
- it passes through the air by convection currents
- it is then conducted through the roof tiles.

To reduce the energy flow through the ceiling, loft insulation in the form of fibreglass is placed above the plasterboard. Fibreglass traps air, which is a good insulator, so less energy is allowed to flow into the airspace to form convection currents in the loft.

Reducing the energy loss from a warm building makes it more **energy-efficient** – less energy needs to be supplied to maintain a comfortable environmental temperature.

Much of the insulation in a building relies on trapped air. We also use trapped air to insulate our bodies in cold weather. Tight-fitting clothes trap a layer of air next to the skin. In very cold weather the most effective way to reduce the heat loss is to wear more layers of clothing – each extra layer traps another layer of air and reduces the energy transfer by conduction.

> It costs about $200 to install loft insulation in an average house, compared to about $10000 for double-glazed windows.

PROGRESS CHECK

1. Which method of transfer of thermal energy:
   (a) can transfer energy through a vacuum?
   (b) only occurs in fluids?
   (c) can occur in solids, liquids and gases?
2. Explain how cavity wall insulation reduces the energy lost from a warm house.
3. Suggest why several layers of paper provide effective insulation for takeaway food.

1(a) radiation; (b) convection; (c) conduction; 2. Air is trapped in pockets. This stops convection currents; 3. A layer of air is trapped between the layers of paper. The trapped air is a poor conductor.

Sample IGCSE questions

1.

(a) The diagram shows a vacuum flask that is used to prevent hot liquids from cooling or cold ones from becoming warm. This is accomplished by minimising the transfer of heat between the flask and the surroundings.

- screw top
- silvered on inside
- contents
- vacuum
- outer container

Briefly discuss those features of the flask, that minimise heat transfer by:

(i) Conduction [2]

The inner container touches the outer one only at a few points, at the top and bottom of the flask. Such minimal contact reduces conduction, which needs a pathway for heat to travel.

(ii) Convection [2]

The vacuum between the walls of the flask means that no fluid is present there and so no convection current can occur.

(iii) Radiation [2]

The flask wall is silvered to reduce the effect of a temperature difference between outside and inside environment.

(b) Places that are located near the sea tend to have a more equable climate than those inland – the summers are cooler and the winters are warmer. Explain why this should be the case. [4]

Places near the sea remain cool in summers because water has a high specific heat capacity and hence takes much energy and a longer time to heat up. In winters, it takes longer to cool than the land surface and, therefore has a warming effect. Places inland have no such moderating influence.

Sample IGCSE questions

2.

(a) Write down five advantages of using mercury, over water, for use as the thermometric substance in a thermometer: **[5]**

> *It has a low thermal capacity and hence responds to temperature changes quickly.* ✔
>
> *It remains liquid over a large range of temperatures.* ✔
>
> *It is non-corrosive.* ✔
>
> *It is easily visible.* ✔
>
> *It expands markedly for a small temperature rise.* ✔

(b) When a thermometer is placed in crushed ice, the mercury thread is 3 cm above the bulb of a thermometer. When it is placed in boiling water, the thread rises to be 15cm above the bulb.

(i) What is the temperature when the mercury thread is 6cm above the bulb? **[3]**

> *The temperature is 25°C.*

(ii) How long is the mercury thread at a temperature of 80°C?

> *At 80°C, the mercury thread is 12.6 cm above the bulb.*

3. Explain the following:

(i) How is evaporation useful for keeping our body cool in summers? **[4]**

> *On a hot day, or after doing some physical work, the temperature of our body tends to rise and the process of sweating causes moisture to appear on our skin. When this moisture evaporates, the latent heat of vaporisation is supplied by our body. Our body thus loses energy and cools.*

Marks can be awarded for the quality of communication in your answers.

(ii) Which type of coolant should be used in radiators of cars? **[4]**

> *A liquid with high specific heat capacity will be useful because it will withdraw a large amount of heat from the radiator for a small rise in temperature. So, the danger of the overheating of engine is minimised and the coolant will last longer in the radiator as it will not boil away.*

(iii) Why does a thick glass tumbler crack when boiling water is poured into it? **[4]**

Sample IGCSE questions

When boiling water is poured in a thick glass tumbler, the glass on the inner side expands due to rise in temperature. Since glass is a poor conductor, very little heat flows out from the inner wall, and so the glass on the outer surface expands very little. The stresses that are set up in the tumbler cause the glass to crack.

Exam practice questions

1. The graph shows the phase and temperature changes as a block of ice (at minus ten degrees celsius) is heated over a bunsen burner. The various points and regions of the graph are labelled with letters, as shown:

Which letter(s) corresponds to:

(a) the freezing point of water   [1]

(b) the boiling point of water   [1]

(c) the melting point of ice?   [1]

Why are the regions BC and DE flat, indicating no temperature change? What is the scientific term used for the parts of the graph corresponding to these regions?   [5]

2. A quantity of hydrogen gas is kept in a cylinder that is fitted with a moveable piston. At a temperature of 27°C, the gas exerts a pressure of 1.2 N/m² and occupies a volume of 500 cm³. Slowly, the piston is pushed downwards so that the gas is compressed to a volume of 200 cm³, the temperature remaining constant.

(a) Express 27°C on the Kelvin scale.   [1]

(b) Calculate the new pressure of the gas after compression.   [1]

The compressed gas is now heated to the boiling point of water.

(c) Write down the temperature, in Kelvin, at which water boils.   [1]

(d) During the heating process, the piston is seen to be pushed backwards. Describe, in molecular terms, the reason for this occurence.   [5]

Exam practice questions

3. The diagram represents the particles of a gas in a cylinder.

(a) Explain how the gas exerts a pressure. [2]

(b) Explain how the gas pressure changes when the temperature of the gas is increased. [3]

(c) The pressure of the gas in the cylinder is 3.5×10^5 Pa when the temperature is 20°C. Calculate the pressure of the gas at a temperature of 50°C. [3]

(d) A label on the cylinder states that "this product must not be left in full sunlight". Explain why this could be dangerous. [3]

4. A kettle is filled with 1.50 kg of water at a temperature of 10°C.

The kettle is switched on until the water boils at 100°C.

(a) Calculate the energy absorbed by the water.

The specific heat capacity of water = 4.2×10^3 J/kg °C. [3]

(b) Give three reasons why the kettle element needs to supply more energy than the answer to **(a)**. [3]

3 Properties of waves

The following topics are covered in this section:

- Wave properties and sound ● Light and the electromagnetic spectrum
- More about refractions and lenses ● Resonance ● Wave interference
- Communicating with waves ● Colour ● The restless Earth

3.1 Wave properties and sound

LEARNING SUMMARY

After studying this section you should be able to:

- *recall and use the wave equation*
- *explain how echoes are used to measure distance*
- *describe the effects of refraction and diffraction.*

What is a wave?

A **wave** is a **vibration** or **oscillation**, a to-and-fro motion, which is transmitted through a material or through space. Waves can transfer **energy** and **information** from one place to another without the transfer of physical material.

There are some properties of waves that are common to different types of wave such as sound, light and radio waves. But there are also ways in which these waves behave differently.

> If you write a letter, paper moves from you to the recipient, but this is not the case if you give the same information over the telephone.

Waves can be classified as either **longitudinal** or **transverse**, depending on the direction of the vibrations compared to the direction of wave travel:

- in a **longitudinal** wave, the vibrations are **parallel** to, or along, the direction of wave travel

- in a **transverse** wave, the vibrations are **perpendicular**, or at right angles, to the direction of wave travel

> Waves that travel through the body of a liquid are longitudinal but surface water waves can be considered to be transverse.

- sound waves are **longitudinal**; **light** and other electromagnetic waves (see section 3.2) are **transverse**.

The diagram, **Fig. 3.1**, shows the vibrations in a longitudinal and a transverse wave.

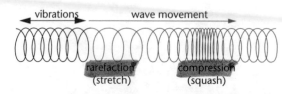

the vibrations in a logitudinal wave (above) and a transverse wave (below)

Fig. 3.1

Wave properties

Wave measurements and the wave equation

These measurements apply to all waves:

- the **amplitude** (symbol a) of a wave motion is the greatest displacement (change in position) from the rest position

- **wavelength** (symbol λ) is the length of one complete cycle of a wave motion – a squash and a stretch for a longitudinal wave, a peak and a trough for a transverse wave

> A common error at IGCSE is to describe the amplitude of a transverse wave as the distance from the top of a peak to the bottom of a trough; it is actually half that distance.

- the **frequency** of a wave (symbol f) is the number of vibrations each second; frequency is measured in hertz (Hz).

> The relationship between the frequency, f, and the time for one oscillation, T, is $f = 1/T$.

Increasing the **frequency** of a wave causes a decrease in the **wavelength**. High-frequency waves have short wavelengths, and low-frequency waves have long wavelengths. Frequency and wavelength are related to wave speed by the **wave equation**.

> **KEY POINT**
>
> For all waves, the relationship between wavelength and frequency is:
> $$\text{speed} = \text{frequency} \times \text{wavelength}$$
> $$v = f \times \lambda$$

The **pitch** and **loudness** of a sound are determined by the frequency and amplitude of the wave:

> Because of the way in which our ears respond to sounds, changing the frequency of a wave can also change the perceived loudness, even though there is no change in amplitude.

- increasing the **amplitude** increases the **loudness** of a sound; a high-amplitude wave sounds louder than one with a lower amplitude

- increasing the **frequency** increases the **pitch** of a sound; a high-frequency sound wave has a higher pitch than a low-frequency wave

- humans can detect sound within the frequency range 20 Hz to 20 000 Hz, but the upper limit is reduced with increasing age

- **compression** waves above the maximum frequency that humans can detect are called **ultrasound**.

Like all other waves except electromagnetic waves, sound needs a medium through which it can travel as it cannot travel through a vacuum. This can be shown by having an electric bell ring inside a glass jar from which the air is being evacuated by a vacuum pump. As the air is removed, the sound of the bell becomes fainter; as we approach vacuum, it ceases altogether.

Sound cannot travel on the moon where there is no atmosphere. The astronomers use an electronic communication device or, as a last resort, touch helmets together to give the sound of their voices some medium to pass through (although the insulation in the helmets will most probably not allow any transference of sound). In air, sound travels at approx. 300 m/s (or 1 km in 3 seconds) although this varies, depending on the day's temperature, pressure and humidity.

This means that:

- if you see a woodcutter chopping a tree in a distant field, you will see him chop the tree and lift his axe again (as light travels enormously quickly) before hearing the sound, which travels much more slowly

- if you look at an aircraft flying overhead, the sound of its engines will come from a point behind the aircraft

- you will hear thunder some time after you see a lightning flash.

A good, and surprisingly accurate, way of finding the speed of sound in air is to strike two pieces of wood together some distance (between 60 and 100 metres) from a large wall. With no other obstacles intervening, an echo can be heard. If the rate of striking is synchronised with hearing the echo, the rhythmic intervals between successive strikes is equal to the time taken by the sound to travel to the wall and back. The distance can be measured and doubled, and then divided by the time taken to calculate the speed of sound.

As a general rule sounds travel quickly through solids (notably metals), less quickly through liquids and slowest through gases. If you put your ear against a table and someone taps the other end, the sound is heard quite distinctly. The effect is best heard over a long distance on railway lines, but do make sure that no train is coming!

Wave reflections and echoes

Echoes of sound and images in mirrors are due to the **reflection** of waves at a surface. **Fig. 3.2** shows the reflection of water waves in a ripple tank.

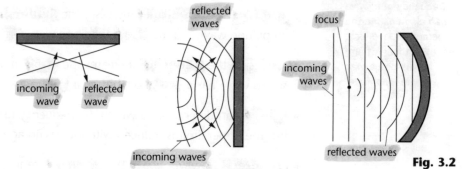

Fig. 3.2

These diagrams in **Fig. 3.2** show that:

- **plane** (straight) **waves** bounce off a barrier at the same angle as they hit it, or "the **angle of incidence** is equal to the **angle of reflection**"

> When you look at your own image in a mirror, it appears to be behind the mirror (see section 3.2).

- **circular** waves from a point source are reflected as if they came from a point behind the barrier; this point is the position of the **image**, the point where the reflected waves appear to have come from

- when plane waves are reflected at a concave barrier they are brought to a **focus**; this is what happens in a satellite dish.

Reflections of sound are called **echoes**. Echoes of sound and ultrasound are used for measuring distances and producing images of the inside of the body.

The diagram, **Fig. 3.3**, shows ultrasound being used to measure the depth of the sea bed.

When used to measure distance:

Fig. 3.3

- a pulse of ultrasound is emitted from a vibrating crystal

- the same crystal then detects the reflected pulse

- since the pulse has travelled from the source, to the object and back again, the distance is calculated as $\frac{1}{2} \times$ speed $\times$ time.

> The pulse of ultrasound travels twice the distance between the source and the object being detected, since it goes there and back again.

ultrasound pulse

reflection

Ultrasound is also reflected at boundaries between layers of different materials and body tissue. **Ultrasound scans** are used to examine railway lines to detect cracks. They are used by medical staff to look at internal organs and fetuses without the need for any incision, so there are no scars to heal. When examining delicate organs and fetuses, an ultrasound scan is safer than an X-ray picture because ultrasound causes no damage to body cells or DNA.

Refraction

Waves change speed when they travel from one material into another or when there is a change in density of the material that they are travelling in. This change of speed causes a change in direction. The change in direction is called **refraction**. The diagram, **Fig. 3.4**, shows the refraction of water waves in a ripple tank.

> In these diagrams, the waves slow down as they pass from deep water into shallow water.

Using a ripple tank

Refraction causes waves to change direction **Fig. 3.4**

When waves are **refracted**:

- the reduction in speed causes a corresponding reduction in wavelength

- the frequency of the waves does not change

- the direction of wave travel does not change if it is at right angles to the boundary between the surfaces

- if the waves meet the boundary at any other angle there is a change in direction.

A swimming pool always looks to be shallower than it really is. This is because the refraction of light as it crosses the water-air boundary causes a **virtual image** of the swimming pool floor to be formed. The diagram, **Fig. 3.5**, shows how the change in direction of light crossing this boundary deceives the eye about the position of the swimming pool floor.

Fig. 3.5

virtual image; this is where the light appears to have come from

> A virtual image is one that light does not pass through and cannot be projected onto a screen. Mirrors form virtual images.

Diffraction

The spreading out of waves when they pass an obstacle or through a gap is called **diffraction**. Diffraction explains how **sound** can be heard around a corner and spreads along a corridor through an open doorway. Diffraction of **light** is less observable than that of sound because of the vast difference in their wavelengths. The amount of spreading of a wave when it is diffracted depends on the relative sizes of the gap that the wave passes through and the wavelength of the waves. This is shown in the diagram, **Fig. 3.6**.

These diagrams shows that:

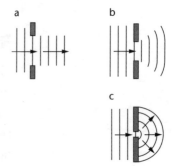

Fig. 3.6

- there is no detectable diffraction when the gap is many wavelengths wide, as in diagram **a**

- there is some spreading of a wave after it passes through a gap that is several wavelengths wide, as in diagram **b**

- after passing through a gap which is one wavelength wide, as in diagram **c**, the maximum spreading of the wave occurs, it appears to originate from the gap.

> When answering questions about diffraction, it is important to emphasise that the amount of spreading of a wave depends on the size of the gap compared to the wavelength.

If you throw a tennis ball against a flat wall, it bounces off at the same angle as it hits. If you try to throw it through a gap, it either goes through or it doesn't, depending on whether the gap is larger or smaller than the ball (and your aim!). It does not spread out after passing through the gap. Both particles and waves show similar behaviour when they are reflected at a flat surface but particles cannot be diffracted. The fact that sound and light can both be diffracted shows that they have a wave-like behaviour.

PROGRESS CHECK

1. Sound travels in air at a speed of 330 m/s.
    Calculate the frequency of a sound that has a wavelength of 0.50 m.
2. SONAR uses sound to measure distances under the surface of the sea.
    A submarine sends out a sound pulse and receives the echo from a ship after 4.50 s.
    The speed of sound in sea water is 1500 m/s.
    How far away is the ship?
3. What is the condition for a wave to spread out in all directions after passing through a gap?

1. 660 Hz; 2. 3375 m; 3. The gap should be the same size as the wavelength of the waves.

3.2 Light and the electro-magnetic spectrum

LEARNING SUMMARY

After studying this section you should be able to:

● *explain how the reflection and refraction of light can lead to image formation*

● *describe how total internal reflection is used in reflecting prisms and optical fibres*

● *recall the main parts of the electromagnetic spectrum and their uses.*

Images from light

When light is reflected at a mirror, or partially reflected by a sheet of glass or a flat water surface, an image is formed. This image is:

● **virtual** – it does not exist but is seen where the brain "thinks" it is

● **directly behind the mirror** the same distance behind the mirror as the object is in front

● **upright** (not inverted)

● the same size as the object.

> When light is reflected by a smooth, flat surface, it bounces off at the same angle as it hits.

An image is seen because the eye-brain system assumes that light travels in straight lines, and uses this principle to determine where the light has come from. This is shown in the diagram, **Fig. 3.7**.

these rays should trace back to the lamp filament

these rays should trace back to the image

Fig. 3.7

The phenomenon of light travelling in straight lines is known as the **rectilinear propagation of light**.

The laws of reflection are as follows:

Note that both angle of incidence and angle of reflection are not angles between the rays and the mirror; they are angles between the rays and the normal to the mirror.

1   The incident ray, reflected ray and the normal at the point of incidence all lie in the same plane (that is, they would lie on the surface of an imaginary sheet of paper).

2   The angle of incidence equals the angle of reflection.

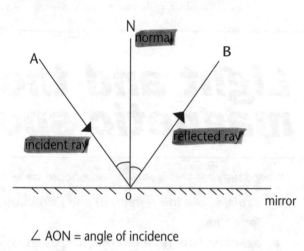

∠ AON = angle of incidence

∠ NOB = angle of reflection   **Fig. 3.8**

PROGRESS CHECK

1.   A man is standing 10 metres away from a plane mirror. He walks 4 m towards the mirror. How far away is he from his image?
2.   A light ray is incident on a plane mirror so that it makes an angle of 25° with the mirror. What is the angle of reflection?

1. 12 metres; 2. 65°.

The bottom of a swimming bath always looks to be nearer than it really is because of the image formed by refraction of light as it passes from water into air.

Light also forms virtual images when it is **refracted** which involves change in speed and direction of light as it passes from one material into another. Light is slowed down when it passes into glass or perspex, and it speeds up again as it emerges. The changes in direction caused by the changes in speed fool the eye-brain system when it works out where the light has come from, so we "see" objects as being closer than they really are.

When light is refracted:

● there is always some reflection at the boundary between two materials

● any change in direction is towards a line drawn at right angles to the surface (this is called a normal line) when light slows down, and away from this line when light speeds up

● light emerges from a parallel-sided block parallel to, but displaced from, the direction in which it entered.

The change in direction of light passing through a glass block and the consequent image formation are shown in the diagrams, **Fig. 3.9**.

Fig. 3.9

> The change in direction when light enters and leaves water is less than that for glass because the change in speed is less.

Notice that, in the above diagrams, light bends towards the normal when it passes from air to glass – this is generally true when light passes from a rarer medium to a denser one. If we measure the sine of the angle of incidence ($i°$) and the sine of the angle of refraction ($r°$) (remember that these angles are not made with the surface but with the normal to the surface), you will find that when we carry out the calculation $\sin i° \div \sin r°$, we always get the same value. This value is called the **refractive index** (n) of the material.

So, for any material, the refractive index,

$$n = \frac{\sin i°}{\sin r°}$$

where i is the angle of incidence on the denser surface and r in the angle of refraction inside it.

KEY POINT

Remember: When light passes into a denser medium, it bends towards the normal.
When light passes into a rarer medium, it bends away from the normal.

This explains how **mirages** occur. In the desert, sunlight passes through the cool atmosphere into hotter layers of air as it approaches the hot sand. Hence, it bends away from the normal. The light that enters the observer's eye comes from the sky, but it appears to have originated from the desert floor as we see in straight lines. Also, as the hot sand heats the air close to it; the air becomes less dense and rises, causing a peculiar shimmering effect.

Internal reflection

> This cannot happen at a boundary where light slows down, for example when light passes from air into glass or water.

Light does not always pass between two transparent materials when it meets the boundary between them. In cases where light would speed up as it crosses a boundary it is possible for all the light to be reflected. The proportions of the light that are **reflected** and **refracted** depend on the **angle of incidence** and the angle between the light that meets the boundary and the **normal** line. The behaviour of light meeting a glass-air boundary at different angles of incidence is shown in the diagram, **Fig. 3.10**.

Fig. 3.10

When light meets a glass-air boundary:

- at a small angle of incidence, some light is reflected and some is refracted
- as the angle of incidence is increased, a greater proportion of the light is reflected
- at the **critical angle** (which is about 42°) the light that leaves the glass is parallel to the surface
- at angles of incidence greater than the critical angle, **total internal reflection** takes place; no light leaves the glass as all of it is reflected internally.

Total internal reflection is used in reflecting **prisms** and to transmit data along **optical fibres**.

Reflecting prisms are used in periscopes, binoculars, cycle reflectors and some cameras. The reflecting prisms in periscopes turn the light round a 90° corner; in binoculars and cycle reflectors the light undergoes two reflections, resulting in a 180° change in direction. This is shown in the diagram, **Fig. 3.11**. In each case the internal angles of the prism are 45°, 45° and 90° so that whenever light meets a glass-air surface the angle of incidence is 45°, which is **greater** than the **critical angle**.

> Reflecting prisms are preferred to mirrors in periscopes because there is no silvered surface that can deteriorate and they absorb less of the light.

Fig. 3.11

> One advantage of optical fibres is that light can travel along them with very little energy loss.

Light can travel "round the bend" in an **optical fibre** by repeated reflection at the glass-air boundary. If the light meets this boundary at an angle of incidence which is greater than the critical angle, all of it is **reflected internally** and none passes out of the fibre.
This is shown in the diagram, **Fig. 3.12**.

An **endoscope** uses two bundles of fibres to see inside the body of a patient. One bundle carries light into the body to illuminate the area being examined. A second bundle carries the reflected light to a television camera, which produces an image on a television screen.

Fig. 3.12

The reason why a diamond sparkles much more than a piece of glass is due to its high refractive index and, therefore, low critical angle. This low critical angle means that much of the light that enters the diamond is trapped within it, bouncing around from facet to facet. The skilled diamond cutter will spend much time examining the diamond for the perfect 'cut', that will enable it to retain as much light as possible.

A spectrum of waves

White light consists of waves with a range of **wavelengths** and **frequencies**. It is split up into the colours of the rainbow when it passes through water droplets. A glass or perspex **prism** can have a similar effect:

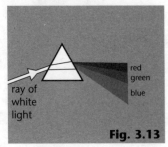
Fig. 3.13

- when light passes into the prism, light at the blue (short wavelength) end of the spectrum undergoes a greater change in speed than light at the red (long wavelength) end of the spectrum

> The range of wavelengths of light in air is approximately 4×10^{-7} m (blue) to 6.5×10^{-7} m (red).

- the change in direction depends on the change in speed

- a similar effect occurs when light leaves the prism

- this splitting up of white light into colours is called **dispersion**.

The dispersion of white light by a prism is shown in the diagram, **Fig. 3.13**.

Hence, the dispersion of light is the phenomenon of splitting of white light into its constituent colours on passing through a prism. The band of seven colours so obtained is called the **visible** spectrum the colours seen are red, orange, yellow, green, blue, indigo, violet.

The recombination of spectrum colours to give white light

First prism splits the incident white light into seven colours. The second prism recombines them to give back the initial white light.

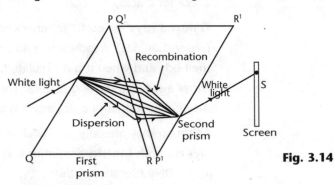
Fig. 3.14

A **rainbow** is the colour effect produced as a result of dispersion of light by rain drops.

Very rarely, a double rainbow can be seen:

- **the primary rainbow** : this has violet colour on its inner edge and red colour on the outer edge

- **the secondary rainbow** : it has red colour on the inner edge and violet colour on the outer edge.

Visible light is only a small part of a whole family of waves with similar properties. All these waves are called **electromagnetic**; this describes the type of oscillation that makes the wave motion. The whole family is collectively referred to as the **electromagnetic spectrum**. It ranges from the shortest electromagnetic waves, **X-rays** and **gamma rays**, to the longest, **radio waves**. The diagram, **Fig. 3.15**, shows the range of wavelengths and frequencies of the different waves that make up the spectrum.

Different types of electromagnetic waves are produced in different ways and have different effects.

frequency/Hz	10^{20}		10^{17}		10^{14}		10^{11}		10^8		10^5
		gamma rays		ultraviolet		infra-red				radio waves	
			X-rays			light		microwaves			
wavelength/m	10^{-12}		10^{-9}		10^{-6}		10^{-3}		1		10^3

Fig. 3.15

All electromagnetic waves:

● travel at the same speed in a vacuum, 3.0×10^8 m/s

● transfer energy and cause heating when they are absorbed.

The shortest waves

X-rays and **gamma rays** have the shortest wavelengths and carry the most energy. They are also the most **penetrative** of the electromagnetic waves, a property which is useful in medical imaging. When an X-ray photograph is taken:

● X-rays are passed through the body and detected by a **photographic plate**

● the X-rays pass through the flesh and are absorbed by the bone

● bone shows up as white on the photograph and flesh appears dark

● a bone fracture is seen as a dark line on the white bone in an X-ray photograph.

A similar technique is used to examine the turbine blades of jet engines, to check for cracks.

Gamma rays are useful for checking individual organs when used as a **tracer** (see section 5.2). A radioactive isotope that emits gamma rays is injected into the body and when it has circulated, it can be detected by a camera to give either a still or a moving picture. The radioactive isotopes can be made so that they concentrate in particular areas of the body.

Although they are only weakly absorbed by body tissue, X-rays and gamma rays are both **ionising** radiations. They can destroy cells and cause **mutations** in the DNA. Because of this:

● both are used in the treatment of cancer to destroy abnormal cells

● gamma rays are used to kill bacteria in food and to sterilise medical instruments

● people who come into contact with X-rays and gamma rays need to be protected from damaging over-exposure.

Either side of light

Ultraviolet radiation is **higher energy** and shorter wavelength than light. Much of the ultraviolet radiation from the Sun is absorbed by the atmosphere, but less amount of this radiation is absorbed in summers than winters. Ultraviolet radiation is also produced when an electric current passes in a tube containing mercury vapour. These tubes are used in sunbeds and in fluorescent lights, where a coating on the inside of the tube absorbs the ultraviolet radiation and re-emits it as light. Security pens use a fluorescent paint which is hardly visible in normal lighting but glows brightly when illuminated with an ultraviolet lamp.

X-rays and gamma rays are produced in different ways. X-rays come from X-ray tubes and gamma rays are emitted by unstable nuclei. There is no difference in the waves.

Due to the Earth's tilt, radiation from the Sun passes through more of the Earth's atmosphere in winter than in summer.

Much shorter wavelength waves, **microwaves**, are used to send the information from London to regional transmitters. These radio waves travel as a narrow beam and so short-wavelength radio waves are used to minimise the effects of **diffraction**. The wavelengths used are typically a few centimetres.

Diffraction effects are even more important when communicating with satellites. The waves used here have wavelengths of a few millimetres. An ordinary television set cannot detect these wavelengths, so to receive satellite television transmissions you also need a set-top box which takes the information from the microwaves and transfers it to longer wavelength waves that can be interpreted by the television set.

> The signals received from satellites are very weak, so a dish aerial is used to gather the energy reaching a wide area.

Television, radio and telephone conversations can be sent as either **analogue** or **digital** signals. The amplitude or frequency of an analogue signal varies continuously with time but a digital signal can only be in one of two states – either "on" or "off". The diagram, **Fig. 3.16**, shows a digital and an analogue signal.

Both analogue and digital signals become **distorted** as they travel in any medium. Digital signals allow higher quality transmission because:

> Digital signals are increasingly being used for all types of communications.

● a digital signal can be returned to its original state in a process known as **regeneration**

● this is possible because any noise or distortion can be detected and removed

● in an analogue signal there is no way of distinguishing between noise and the original signal, and when the signal is amplified so is the noise.

an analogue signal

a digital signal

Fig. 3.16

Digital signals are also used to transmit data along **optical fibres**. Light has a higher frequency than radio waves, allowing the transfer of more information. The diagram, **Fig. 3.17**, shows how optical fibres are used in telephone links.

copper wire

copper wire

optical fibre

telephone exchange

telephone exchange

Fig. 3.17

PROGRESS CHECK

1. What causes the change in direction that can occur when light passes from one material into another?
2. Which three types of wave in the electromagnetic spectrum have wavelengths shorter than light?
3. Why are microwaves used for satellite communication?

1. The change in speed; 2. Ultraviolet radiation, X-rays and gamma rays; 3. To minimise the effects of diffraction.

In addition to these uses, ultraviolet radiation can be harmful to humans:

- their absorption by the skin can cause cancer
- light skins are more prone to cancer than dark skins, as they allow the radiation to penetrate further into the body
- their absorption by the **retina** can cause **blindness**.

> It is important to protect the skin and eyes from ultraviolet radiation when outside in the summer months.

Infra-red radiation has a longer wavelength and **lower energy** than light, so it is less harmful than ultraviolet radiation. All objects **emit** infra-red radiation and **absorb** it from their surroundings. The **hotter** the object, the **greater** the rate of emission. Infra-red radiation is used:

- in cooking; toasters and grills transfer energy to food by infra-red radiation
- in night-time photography; people and animals can be distinguished from their surroundings because they emit more infra-red radiation
- to find and rescue people trapped in the rubble of buildings after an earthquake
- in remote controls for devices such as televisions and hi-fi.

Overexposure to infra-red radiation from the Sun causes **sunburn**.

> You cannot tune in a radio to a microwave oven as the waves do not carry a signal.

The **microwaves** used in cooking have a wavelength around 12 cm, so they fit into the **radio wave** part of the electromagnetic spectrum. Radio waves of this wavelength can penetrate a few centimetres into food, and they have the right frequency to be **absorbed** by **water molecules** as they pass through. Food becomes cooked by the following process:

- water molecules absorb the energy of the microwaves, increasing their energy of **vibration**
- this energy is transferred to other molecules in the food by **conduction**
- food is cooked uniformly as this process takes place throughout the body of the food.

Communicating with electromagnetic waves

National television and radio programmes are broadcast using **radio waves**. Television broadcasts from your local transmitter to your aerial use wavelengths around 0.6 m; the wavelengths used for radio range from around 3 m for VHF to hundreds of metres for medium and long-wave broadcasts. Long wavelength radio waves:

> The signal is carried by changes in either the amplitude or frequency of the radio wave. This is known as amplitude modulation (AM) or frequency modulation (FM).

- are **diffracted** more around hills and buildings, so there is less chance of a "shadow" causing poor reception
- have a **low frequency** which limits the amount of information that can be transmitted.

3.3 More about refraction and lenses

LEARNING SUMMARY

After studying this section you should be able to:

● *describe the action of a lens*
● *draw a diagram to determine the properties of an image formed by a lens*
● *explain how a lens is used in devices such as a magnifying glass and a camera.*

Refractive index

When light travels from one transparent material into another, there is a change of speed. Unless the light is incident along the normal line, this causes a change in direction. The greater the change in speed, the greater the change in direction at a given angle of incidence.

The **refractive index** of a material is a measure of the change in speed that occurs when light passes from a vacuum (or air) into that material.

> There is very little difference between the speed of light in a vacuum and in air, so the refractive index can be taken as the ratio of the speed of light in air to the speed in the material.

KEY POINT

$$\text{refractive index} = \frac{\text{speed of light in vacuum}}{\text{speed of light in material}}$$

The greater the refractive index, the greater the change in speed that occurs. The refractive index of water is around 1.3 and that of glass 1.5. When light passes from air into glass, there is a greater change of speed than when it passes from air into water.

In a vacuum, light of all wavelengths travels at the same speed. This is not the case when light travels in a material such as water or glass where:

● the shorter the wavelength, the slower the speed
● short-wavelength waves have a greater refractive index and a greater change in speed than longer-wavelength waves as they pass from air into a transparent material
● the change in direction depends on the change in speed, resulting in **dispersion** of white light into different colours.

> Dispersion as light passes through water droplets is the cause of a rainbow.

The dispersion of white light as it passes into glass is shown in the diagram, **Fig. 3.18**.

> This diagram shows that red light is deviated the least when it passes from air into another transparent material, with blue light being deviated the most.

Fig. 3.18

The action of a lens

When light is emitted from or reflected by an object, it spreads out or **diverges**. The effect of a **concave**, or **diverging**, lens is to increase this divergence while a **convex**, or **converging**, lens decreases it. The diagram, **Fig. 3.19**, shows the shape of these lenses and their action on light.

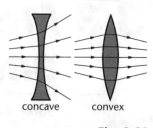

concave   convex

Fig. 3.19

The **principal focus** or **focal point** of a lens is the point on the axis through the centre of the lens where:

● the light converges in the case of a convex lens

● the light appears to diverge from in the case of a concave lens.

> **KEY POINT** The focal length, f, of a lens is the distance from the centre of the lens to the principal focus.

This is shown in the diagram below, **Fig. 3.20**.

> A convex lens does not always cause light to converge, in some cases it just makes it diverge less.

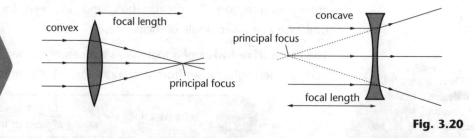

Fig. 3.20

Finding the image

Lenses are used to produce images in devices such as telescopes, microscopes and cameras. An image can be **real** or **virtual**. A real image is one that light actually passes through and can be displayed on a screen, for example the image formed on the film in a camera. No light passes through a virtual image and it cannot be displayed on a screen – it only appears to be there. The image that you see when looking in a mirror is a virtual image.

> Images caused by the refraction of light as it passes through a water-air boundary are virtual – an example is the bottom of a swimming pool; it always looks to be nearer than it really is.

Whether an image is real or virtual, and its position and size, can be found by drawing a **ray diagram**. This relies on two key facts:

● light passes through the centre of a lens undeviated

● light incident parallel to the axis passes through the principal focus in the case of a convex lens and appears to have come from the principal focus in the case of a concave lens.

> It is easier to draw ray diagrams on squared paper such as graph paper rather than lined or plain paper. The grid enables you to set up a scale and draw parallel lines.

The procedure for drawing a ray diagram is as follows:

● draw a vertical line to represent the lens and a horizontal line to represent the axis through the lens

● to a suitable horizontal scale, draw an upright arrow on the axis to represent the object

- using the same scale, mark the position of the principal focus on the axis. This should be placed on the same side of the lens as the object for a concave lens, and on the opposite side for a convex lens.
- Draw a line from the top of the arrow, parallel to the axis until it meets the lens. Continue this line through the principal focus (for a convex lens) or as if it had come from the principal focus (for a concave lens).
- Draw a second line from the top of the arrow, straight through the centre of the lens.

If these lines cross after passing through the lens, the image formed is **real**. The top of the image is where the lines cross, and the bottom of the image is the point directly below or above it on the axis. If the lines do not cross, the image is **virtual**. In this case the image position is located by tracing the lines back, to find the point where they would both appear to have come from to an observer looking at the light that has passed through the lens.

The position and size of the image can be judged from the horizontal scale. The diagram, **Fig. 3.21**, shows completed ray diagrams for a convex and a concave lens.

> Dotted lines are used here for the path that light appears to have followed and to represent a virtual image.

convex lens   concave lens

Fig. 3.21

Types of image

A concave lens on its own always forms an image that is virtual, upright and smaller than the object. Convex lenses can form three types of image, depending on the distance between the object and the lens:

> If the object is placed precisely at f, no image is formed – the light that emerges from the lens is parallel.
>
> If the object is placed precisely at $2f$, the image is real, inverted and the same size as the object.

- if the object is between the lens and a distance equal to its focal length, f, the image is virtual, upright and magnified – this corresponds to using the lens as a **magnifying glass**
- if the object is between f and $2f$ from the lens the image is real, upside down and magnified – this corresponds to using the lens as a **projector**
- if the object is further away than $2f$, the image is real, upside down and smaller than the object – this corresponds to using the lens as a **camera**.

Table 3.1 Position of objects and images formed by a convex lens

Position of object	Position of image	Size of image	Nature of image
Between F and lens	On the same side as object	Enlarged	Virtual and erect
At focus (F)	At infinity	Highly enlarged	Real and inverted
Between F and 2F	Beyond 2F	Enlarged	Real and inverted
At 2F	At 2F	Same size as object	Real and inverted
Beyond 2F	Between F and 2F	Diminished	Real and inverted
At infinity	At focus (F)	Highly diminished	Real and inverted

Optical instruments

To use a convex lens as a **magnifying glass**, you should look through the lens at the object. The diagram, **Fig. 3.22**, shows that the light that has passed through the lens is less divergent than when it entered the lens. This produces a magnified image that is further away than the object. The closer the object is to the principal focus, the greater the magnification.

object

virtual, magnified image

eye placed here

Fig. 3.22

A **projector** and a **camera** both produce real images; that from the projector is magnified but the image that forms on the film in a camera is usually smaller than the object being photographed. The diagrams, **Fig. 3.23**, show the action of a projector and a camera.

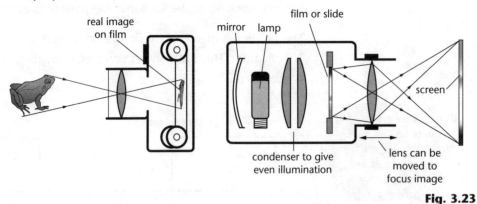

real image on film

film or slide

mirror   lamp

screen

condenser to give even illumination

lens can be moved to focus image

Fig. 3.23

To focus these instruments and make the image on the screen sharper:

* the camera lens is moved away from the film to focus on a near object, and towards the film to focus on a close object
* the projector lens is moved towards the slide to increase the magnification when the screen is moved away from the projector.

Sight

The eye uses two convex lenses, one fixed and one variable, to produce a real image. The diagram, **Fig. 3.24**, shows some key components of the eye.

> The fixed lens is the cornea, which does most of the focusing. This is then "fine-tuned" by the eye lens.

ciliary muscles

iris

pupil

lens

cornea

retina

optic nerve

Fig. 3.24

The cells that are sensitive to brightness are called rods and those sensitive to colour are called cones. Cones do not function in low light levels, so you cannot tell the colour of grass by moonlight!

- the **retina** contains cells that are sensitive to the brightness and colour of light
- electrical impulses from these cells are sent along the **optic nerve** to the brain
- the **iris** is a circular piece of tissue that controls the size of the **pupil**
- the **ciliary muscles** adjust the shape of the **lens** to focus on near and distant objects; this process is called **accommodation**.

Two common reasons for wearing spectacles or contact lenses are:

- **short sight**, where near objects can be seen clearly but the light from a distant object forms a real image in front of the retina
- **long sight**, where distant objects can be seen clearly but light from a near object is not brought to a focus in the eyeball.

The diagram, **Fig. 3.25**, shows the effects of short and long sight and how they can be corrected using lenses.

short sight   and its correction   long sight   and its correction

light from a distant object

light from a near object

Fig. 3.25

PROGRESS CHECK

1. The speed of light in air is 3.00×10^8 m/s.
   The refractive index of glass is 1.50.
   Calculate the speed of light in glass.
2. Where should an object be placed, relative to the lens, when a convex lens is used as a magnifying glass?
3. A 2 cm × 2 cm slide is used in a slide projector.
   The slide is placed 12.0 cm from the projection lens, which has a focal length of 10.0 cm. Use a ray diagram to find how far the screen should be placed from the lens.

1. 2.00×10^8 m/s;   2. Between the lens and its principal focus;   3. 60 cm.

3.4 Resonance

LEARNING SUMMARY

After studying this section you should be able to:

● recall the meaning of natural frequency
● explain how resonance occurs
● describe the modes of vibration of a string.

Natural vibrations

Every object has its own **natural frequency** of vibration, known as its **characteristic frequency**. This is the frequency of the oscillation when it is displaced and allowed to vibrate freely. Plucking a rubber band or pushing a swing causes a vibration at the natural frequency.

> The natural frequency of a rubber band can be changed by altering the tension.

> The stiffness of a spring is a measure of how difficult it is to stretch it.

The diagram, **Fig. 3.26**, shows two systems whose natural frequency of vibration can be changed.

The natural frequency of a pendulum depends only on the length of the string, but that of a mass on a string depends on the mass and the stiffness of the spring

Fig. 3.26

All objects can also be forced to vibrate by bringing them in contact with another vibrating object, but in many cases, the amplitude of a **forced vibration** is small.

> **KEY POINT**
> If the frequency of a forced vibration is the same as the natural frequency, then a large amplitude vibration builds up. This is called resonance.

Examples of resonance include:

> Electrical resonance is used to tune radios and televisions to different stations.

● pushing a child on a swing – the amplitude increases when the swing is pushed at a frequency equal to its natural frequency
● rattling exhausts – parts of a motor vehicle can vibrate with a large amplitude when their natural frequency matches that of the engine vibrations
● playing a clarinet – a column of air is forced to vibrate by a vibrating reed.

Blowing across a column of air, such as in an empty or partly-filled milk bottle, can also cause it to vibrate at its natural frequency. This is how a church organ produces notes. Different notes are played by pipes of different sizes – long pipes have lower natural frequencies than shorter ones.

Vibrating strings

Pianos, guitars and violins all use the vibrations of a string in tension to produce sound. Strings can vibrate in a number of ways, called **modes**. The diagram, **Fig. 3.27**, shows the vibrations of a string at its natural frequency. This is known as the **fundamental mode** of vibration.

> In the fundamental mode the length of the vibrating string is half the wavelength of the sound wave produced.

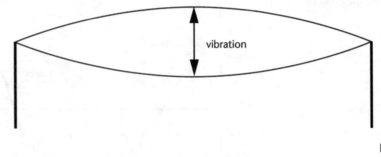

Fig. 3.27

In this mode of vibration:
- the ends of the string are **nodes** – points of no displacement
- there is an **antinode** – where the maximum displacement occurs – in the centre of the string.

> The lowest frequency notes on a piano are produced by the longest strings – these also have a greater mass per unit length than the strings used to play the high notes.

The frequency of the fundamental mode of vibration of a string depends on:
- its length – increasing the length reduces the natural frequency
- the mass per unit length – the greater this is, the lower the natural frequency
- the tension – increasing the tension increases the natural frequency.

Strings can also vibrate with large amplitudes at frequencies equal to a whole number multiple of the natural frequency. The diagram, **Fig. 3.28**, shows the modes of vibration of a string at frequencies equal to twice and three times the natural frequency.

> The fundamental mode of vibration has just one antinode, but other modes have several.

Fig. 3.28

When a stringed instrument is played, the strings vibrate in several modes at once. This is why the same note played on different instruments has a different sound. The **quality** of a note depends on the number and relative intensity of the different modes of vibration. The displacement–time graphs, **Fig. 3.29**, compare the "pure" note from a tuning fork with the same note played on a piano.

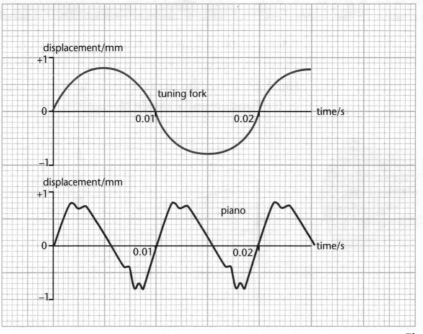

Analysis of the note from a tuning fork shows that it is a sine wave. A signal generator can also be used to produce a "pure" note.

Fig. 3.29

A displacement–time graph can be used to determine:

- the amplitude of the wave – the maximum displacement
- the frequency of the wave by measuring the time, T, for one complete cycle and using the relationship $f = 1/T$
- the quality of the note by judging how much it deviates from a "pure" note.

PROGRESS CHECK

1. Suggest why a washing machine can vibrate with a large amplitude as the speed of the spin increases.
2. What is required for an object to resonate?
3. What affects the quality of the note played on a violin?

1. The vibrations of the drum are at the natural frequency of the washing machine;
2. A forced vibration at the same frequency as the natural frequency; 3. The number and intensity of the modes of vibration present.

3.5 Wave interference

LEARNING SUMMARY

After studying this section you should be able to:

● *recall that interference involves cancellation and reinforcement of waves*
● *describe interference effects with sound, surface water waves and electromagnetic waves*
● *explain how constructive and destructive interference are related to path difference.*

What is interference?

Interference is a property of all wave motions. It results in the **cancellation** or **reinforcement** of waves when the paths of two waves of the same type cross. The effects of interference are readily observed using sound but are more difficult to observe with short-wavelength waves such as light. The diagram, **Fig. 3.30**, shows how an observer can detect the effect of sound waves interfering.

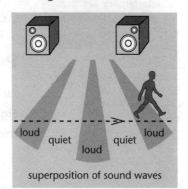

superposition of sound waves

Fig. 3.30

The observer notices that:

● there are places where the sound is louder than that due to one loudspeaker alone
● there are places where the sound is almost inaudible.

> The separation of the loud and quiet sounds depends on three factors; the wavelength of the sound, the separation of the loudspeakers and the distance between the sources and the observer.

This is due to **constructive interference** and **destructive interference** taking place between the waves from the two sources.

 KEY POINT

Constructive interference results in a loud sound. It occurs at any point where the displacements from both waves are in the same direction. Destructive interference occurs at points where the displacements are in opposite directions and results in a quiet sound.

The diagram, **Fig. 3.31**, shows how two waves interfere constructively and destructively when they cross each other.

> The waves only show complete cancellation if they have the same amplitude.

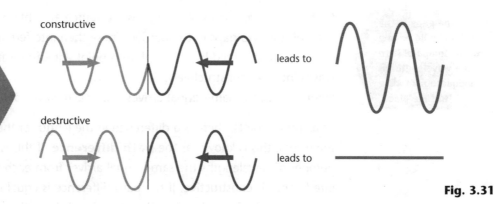

constructive

leads to

destructive

leads to

Fig. 3.31

Interference on water

Interference effects can also be noticed with surface water waves. The diagram, **Fig. 3.32**, shows what happens when two dippers create circular waves on water. The solid lines represent wave crests and the broken lines represent wave troughs.

Fig. 3.32

> The factors that affect the separation of the lines of constructive and destructive interference are the same as those for the sound waves.

Where two wave crests or two wave troughs cross, they interfere constructively – some lines of constructive interference (c) have been marked on the diagram. Destructive interference takes place at points where a wave crest meets a trough. Lines marked (d) show destructive interference.

Path difference

All electromagnetic waves show interference effects also. The diagram, **Fig. 3.33**, shows an arrangement for demonstrating interference of 3 cm microwaves.

Fig. 3.33

> The gaps in the barrier should have a width comparable to the wavelength for adequate spreading to take place.

In this case, two copies of a single wave source are obtained by **diffraction** through the two gaps in the barrier. These then interfere and variations in intensity are detected by an aerial. At a point equidistant from both gaps, the waves interfere constructively. Since the waves have travelled along equal-length paths, the same signal arrives from each wave source.

At all other points, there is a difference in the length of the paths travelled by the waves; this is known as the **path difference**. If this is equal to a whole number of wavelengths the same signal arrives from each wave and the interference is constructive. If the path difference is equal to half a wavelength or one and a half wavelengths, then the signals from the two waves have opposite displacements, hence they interfere destructively.

This is shown in the diagram, **Fig. 3.34**

the path difference here is half a wave length, giving destructive interference

the path difference here is one wave length, giving constructive interference

Fig. 3.34

This is true provided that the two wave sources are vibrating in step with each other.

KEY POINT

Constructive interference occurs when the path difference between waves from two sources is equal to a whole number of wavelengths, or an even number of half wavelengths.

Destructive interference occurs when the path difference is equal to an odd number of half wavelengths.

Interference is more difficult to observe with light due to its shorter wavelength. The fact that it does happen provides evidence for light having a wave-like nature.

PROGRESS CHECK

1. Explain what is meant by *destructive interference*.
2. How is *diffraction* used in the demonstration of interference between two sources of 3 cm electromagnetic waves?
3. What is the condition for waves from two sources vibrating in step to interfere constructively?

1. The cancellation of two waves with opposite displacements;   2. To produce two identical copies of a single wave source;   3. The path difference should be a whole number of wave-lengths, or an even number of half wavelengths.

3.6 Communicating with waves

LEARNING SUMMARY

After studying this section you should be able to:

- describe how information is transmitted as analogue and digital signals
- explain how the transmission of radio waves through the atmosphere depends on the wavelength of the waves
- describe some uses of satellites.

Systems

Communications using sound, light and radio may appear to be very different processes, but there are a number of features common to many communication systems:

- a **transducer** transfers a signal from one form to another
- an **encoder** translates the information into a form that can be transmitted
- a **modulator** produces periodic variations in amplitude and/or frequency of the wave that carries the coded signal

> Sound is modulated by the mouth during speech.

- an **amplifier** increases the amplitude of a signal
- a **transmitter** sends the signal from one place to another
- a **receiver** detects a transmitted signal
- a **demodulator** separates a coded signal from the wave that carries it
- a **decoder** translates the information from the form in which it is transmitted to one in which it can be understood.

Transmitting along wires and fibres

Signals, that carry telephone conversations and radio and television programmes, spend some part of their journey from the transmitter to the receiver travelling along wires or fibres or both. In each case:

> All signals are subject to attenuation as they travel.

- the signals become **attenuated** due to loss of energy – this results in them becoming weaker
- they pick up **noise** which affects the amplitude of the wave and results in distortion of the signal.

> Electrical signals travelling in wires need to be repeated every few miles. Optical signals travelling along fibres have a much greater range.

Most optical fibre transmissions use digital signals, where the information is encoded as a series of pulses of light or infra-red radiation. Due to its higher frequency, light can carry more information that radio waves. Both analogue and digital signals need to be amplified at several stages during their transmission because of the effects of attenuation. Digital signals can be restored to their original condition in a process known as **regeneration** (see section 3.2). Analogue signals are amplified by repeaters, which amplify both the signal and the noise. This is shown in the diagram, **Fig. 3.35**.

an analogue signal

becomes noisy and distorted

which is amplified

a digital signal

becomes noisy and distorted

and can be regenerated

Fig. 3.35

Storing and reading information

Computer disks store information in a **digital** form – each **bit** of information can be a 0 or a 1. This information is stored in a magnetic form. A **compact disc** (CD) also stores information digitally as a series of "pits" and "bumps" on its surface. The diagram, **Fig. 3.36**, shows how a **diode laser** and a **photodiode** are used to read the information on a compact disc.

diode laser

photodiode detector

laser beam

prisms

reflected beam

compact disc

Fig. 3.36

> The photodiode detects light from the diode laser that has been reflected at the surface of the CD.

A **vinyl record** stores information in an **analogue** form – here the information is stored in grooves that have both horizontal and vertical bumps. The bumps vary in frequency and amplitude in the same way as the sound recorded.

> A domain is a magnetised region on the tape.

Magnetic tape can be used to store either type of signal by rearranging the magnetic **domains** on the recording tape. On an unrecorded tape, these are arranged randomly, but when a signal is recorded these domains are ordered in a way that represents the information either as a varying magnetic field for an analogue signal or a series of magnetic pulses in the case of a digital signal.

Information is written onto and read from magnetic tape using a tape head similar to that shown in the diagram, **Fig. 3.37**.

There are three types of tape head in a tape recorder:

a.c. input

soft-iron core

tape

narrow gap filled with non-magnetic material

Fig. 3.37

- the record head writes information onto the tape when an alternating current passes in the coil and produces an alternating magnetic field in the gap that the tape passes over

> In some inexpensive tape recorders the same tape head is used for recording and playback.

- the erase head re-scrambles the magnetic domains when a high frequency alternating current passes in the coil; this removes any recorded information
- the playback head reads information by electromagnetic induction – an alternating voltage is induced in the coil due to the changing magnetic field of the moving tape.

The record and playback heads are examples of **transducers**; they transfer information from one form to another.

Other key transducers

A **microphone** transfers the information from a signal in the form of sound to an electrical signal. One common form of microphone is the **moving coil microphone**, shown in the diagram, **Fig. 3.38**. In this transducer:

diaphragm   coil

Fig. 3.38

- the sound wave causes vibration of the diaphragm
- this causes the coil to vibrate in the radial magnetic field of a cylindrical magnet
- a voltage is induced in the coil with a variation in frequency and amplitude similar to that of the sound.

In a **loudspeaker** the process is reversed – information from an electrical signal is reproduced as a sound. The diagram, **Fig. 3.39**, shows a **moving coil loudspeaker**. In this transducer:

cylindrical magnet

paper cone

Fig. 3.39

- an alternating current that represents the signal passes in the coil
- interaction of the magnetic fields due to the current and the fixed magnet produces a force on the coil
- the coil vibrates, pushing the paper cone in and out.

Using radio waves

Prior to the development of radio, long-range communication depended on electric currents and needed a wire link from the transmitter to the receiver. In the mid-nineteenth century the **telegraph** system was used to send messages using Morse code. Then the invention of the telephone by *Alexander Graham Bell* in 1876, enabled people to be able to talk directly to each other. At the beginning of the twentieth century, radio was beginning to be developed and was the first "wireless" means of long-distance communication. More recent developments include the use of satellites which have made world-wide communications possible.

Radio waves can travel in a number of different ways, depending on their wavelength and frequency:

- long-wavelength waves with frequencies up to 2 MHz travel as **ground waves** – they follow the curvature of the Earth
- medium-wavelength waves with frequencies in the range 3–30 MHz are reflected by the ionosphere, a layer in the upper atmosphere, and the ground – they are called **sky waves**
- short-wavelength waves with frequencies above 30 MHz travel in straight lines and can pass through the ionosphere – these are called **space waves** as they can travel through the atmosphere and out to space.

The diagram, **Fig. 3.40**, shows how these waves travel.

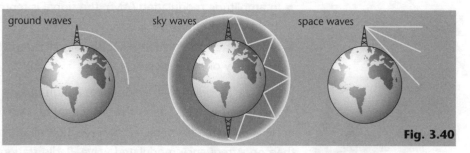

ground waves   sky waves   space waves

Fig. 3.40

By virtue of their range, long-wavelength radio broadcasts are particularly suited for transmission to ships. In Britain, a single transmitter in the middle of the country is used to broadcast a radio station that can be detected all over the country and abroad. Satellite transmissions use short wavelengths that can travel through and beyond the atmosphere.

> Interference can cause both cancellation and reinforcement of a radio signal.

The reception of radio waves is affected by **interference** (see section 3.5) and **diffraction** (see section 3.2). Interference can affect reception in a number of ways:

- interference between a wave that travels directly from a transmitter to a receiver and one that is reflected by the ionosphere can cause a signal to fade and weaken
- interference between a wave that travels directly from a transmitter to a receiver and one that has been reflected from a large wall or building can have a similar effect
- waves given out from adjacent transmitters can interfere, creating some regions where the signal is strong and some other where it is weak; this is usually avoided by using different wavelengths in areas where the signal from more than one transmitter can be received.

Diffraction enables the reception of long-wavelength waves and limits the reception of short-wavelength waves. The amount of spreading of waves as they meet obstacles such as buildings and hills depends on the wavelength; the greater the wavelength, the more the spreading that occurs. The diagram, **Fig. 3.41**, shows how a remote house can receive long-wavelength broadcasts but not short-wavelength ones.

Fig. 3.41

> It is important to minimise diffraction effects to reduce attenuation of the signal.

Diffraction also affects transmission of signals that carry national radio and television broadcasts over long distances. In this case, a narrow beam is required to transmit the signal between dish aerials. To minimise diffraction effects, the wavelength used to transmit the signal should be much smaller than the diameter of the dish. As there is a limit to the size of the dishes that can be used, this dictates the use of short-wavelength waves.

Radio waves carry signals by **modulation** of the wave:

- in amplitude modulation (AM), the signal is represented by changes in amplitude of the wave
- in frequency modulation (FM), the signal is represented by changes in frequency of the wave.

Frequency modulation is used where a high quality of reception is required. This is because noise affects the amplitude of a wave but not the frequency, so noise is more readily detected on reception of an amplitude modulated wave than on a frequency modulated wave. Radio stations in Britain use amplitude modulation for the longer-range medium and long wave broadcasts and frequency modulation for the higher quality VHF (very high frequency) broadcasts.

The diagram, **Fig. 3.42**, shows how a signal is carried by amplitude modulation and by frequency modulation.

> High-quality radio broadcasts need to use very high frequencies because more information can be sent on a high frequency wave than on a low frequency one.

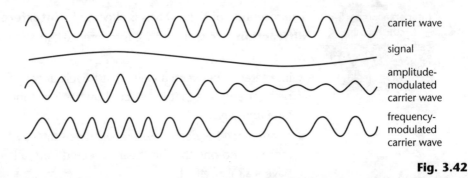

Fig. 3.42

Producing and detecting a radio wave

The essential parts of an AM radio transmitter are:

- an oscillator that produces the high frequency radio waves
- a modulator that changes the amplitude of the radio wave to carry the low-frequency sound signal
- an aerial to broadcast the radio waves.

After the waves have been received by an aerial, the signal is in the form of an electrical oscillation. The sound frequency signal has to be removed from the radio frequency signal in a process known as **detection** before it is amplified so that it can drive a loudspeaker. The diagram, **Fig. 3.43**, shows how a diode is used along with a resistor and a capacitor to extract the sound signal from the amplitude modulated radio signal.

> The radio wave is called the carrier wave because it carries the sound signal.

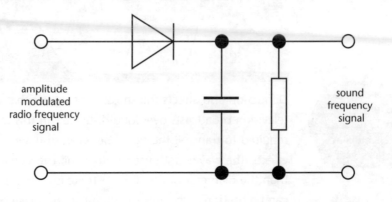

Fig. 3.43

Using satellites

Some early satellites used for a radio transmission merely acted as reflectors of radio waves – they were **passive satellites**. An **active satellite** receives, amplifies and re-transmits signals. Artificial satellites in orbit around the Earth are used for a number of purposes:

Satellite navigation systems can now be used to track the movement of a stolen vehicle.

- **navigation** – a ship or a car can use satellite transmissions to determine its position with a precision of a few metres
- **exploration** of the Solar System and the Universe – satellites such as the Hubble telescope can detect radiation that is absorbed by the Earth's atmosphere
- **surveillance** – the movement of armies and weapons can be monitored by satellites
- **weather** – the movement of clouds and weather fronts is monitored by satellites
- **communications** – telephone conversations and television broadcasts are transmitted around the world by satellites.

The time it takes for a satellite to complete one orbit of the Earth depends on the height of the satellite. Satellites occupy different orbits according to the job that they do.

Some weather satellites occupy **low polar orbits** similar to that shown in the diagram, **Fig. 3.44**. In a low polar orbit:

The Earth rotates through an angle of 24° during one orbit of a satellite in a low polar orbit.

- a satellite completes an orbit of the Earth in 96 minutes
- during this time the Earth spins on its axis, so the satellite "sees" a different part of the Earth on the next orbit
- the whole of the Earth's surface is monitored during one day.

equator night

Fig. 3.44

For telecommunications, satellites that do not move relative to the Earth are often needed. These occupy **geostationary orbits**. A satellite in a geostationary orbit:

- completes one orbit of the Earth in 24 hours
- remains in the same position above the Earth's surface
- orbits above the equator.

PROGRESS CHECK

1. Which part of a domestic radio is the receiver?
2. How is information encoded before being transmitted along an optical fibre?
3. Explain why satellite transmissions use very short wavelength radio waves.

1. The aerial; 2. As pulses of light or infra-red radiation; 3. Long wavelength waves cannot pass through the ionosphere.

3.7 Colour

LEARNING SUMMARY

After studying this section you should be able to:

● *describe the difference between a primary colour and a secondary colour*
● *explain the effect of colour filters*
● *predict the appearance of coloured objects in different colours of light.*

Primary and secondary colours

The visible part of the electromagnetic spectrum consists of an uncountable number of colours ranging from blue to red. In a colour television and computer monitor, all of these colours (and some additional ones) are produced from a combination of just three colours: red, green and blue.

KEY POINT

Red, green and blue are called the primary colours because any other colour of light can be obtained by combinations of these three colours.

> *Mixing coloured lights is colour addition. It is not like mixing paints which, like colour filters, subtract colours.*

The colour obtained by combining two primary-coloured lights of equal intensity is called a **secondary colour**. The diagram, **Fig. 3.45**, shows the results of overlapping beams of light of the three primary colours on a white screen. This diagram shows that:

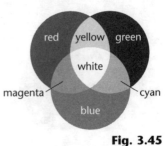

Fig. 3.45

● yellow light is produced by mixing red light and green light
● magenta light is produced by mixing red light and blue light
● cyan light is produced by mixing blue light and green light

> *Secondary colours are used in colour printing and colour photographs.*

● white light is produced by mixing all three primary colours of light.

Colour filters

Colour filters remove some colours from white light, and allow others to pass through. A primary colour filter allows light of its own colour only to pass through, but that of a secondary colour transmits its own colour as well as the two primary colours that combine to make the secondary colour. The table shows the primary colours of light, transmitted and absorbed by different coloured filters.

> *Colour filters are used in theatres to produce dramatic effects.*

Colour of filter	Primary colours transmitted	Primary colours absorbed
red	red	green and blue
green	green	red and blue
blue	blue	red and green
yellow	red and green	blue
magenta	red and blue	green
cyan	green and blue	red

When two filters are combined, each one absorbs certain colours. Hence the light that emerges contains the colours that are not absorbed by either filter. The diagram, **Fig. 3.46**, shows what happens when white light passes through a yellow filter followed by a magenta one.

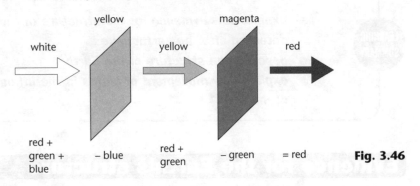

Fig. 3.46

Coloured objects

Most objects are seen by the light that they reflect. A white surface reflects all colours of light and a black surface absorbs all colours and reflects none. Other colours of surface are more selective – some colours are absorbed and some are reflected. The colours that any particular object absorbs and reflects can be deduced from its appearance in white light, since coloured objects act in a similar way to colour filters.

Objects look to be different colours in coloured lights because some of the colours that they normally reflect are not present. To predict what colour an object looks to be when coloured light is shone on it:

● write down the primary colours present in the coloured light
● cross out the primary colours that the object absorbs
● if no primary colours are left, the object appears black. If one primary colour is left it appears to be that colour and if two primary colours are left it appears to be the secondary colour produced when light of those two primary colours is combined.

The diagram, **Fig. 3.47**, shows the effect of illuminating a yellow tee-shirt with magenta coloured light.

Fig. 3.47

1. Which two primary colours of light combine to form (a) cyan, (b) magenta and (c) yellow?
2. Explain why no light passes through a combination of two primary colour filters.
3. A top appears cyan in white light.
   What colour does it appear in yellow light?

PROGRESS CHECK

1a. blue and green; b. red and blue; c. red and green; 2. The primary colour transmitted by the first filter is absorbed by the second; 3. Green.

3.8 The restless Earth

LEARNING SUMMARY

After studying this section you should be able to:

- *explain how evidence for the structure of the Earth comes from the waves detected after an earthquake*
- *describe the structure of the Earth*
- *explain how movement of plates in the lithosphere results in the recycling of rock.*

Evidence for the Earth's structure

Earthquakes produce three types of wave that travel through the Earth and are detected by instruments called **seismometers**. The diagram, **Fig. 3.48**, shows a recording of the waves detected by a seismometer.

> A seismometer recording is called a seismographic record.

Fig. 3.48

The waves detected are:

- **L waves**; these are **long-wavelength** waves that travel around the Earth's crust. They are responsible for damage to buildings caused by movements in the ground.

> The relative speeds of the waves can be deduced from the seismometer recording shown in Fig. 3.48.

- **P waves** or primary waves; these are **longitudinal** waves that can travel through both solids and liquids. They have the greatest speed of the waves caused by an earthquake and are the first to be detected.

- **S waves** or secondary waves; these **transverse** waves can only travel through solid materials. They are detected after the primary waves because of their lower speed.

The diagram, **Fig. 3.49**, shows where P and S waves are detected following an earthquake.

This diagram shows that:

- both P and S-waves follow curved paths due to **refraction** as the speed of the waves increases due to increasing density of the material that they are travelling through

Fig. 3.49

- S waves are not detected in the **shadow region** on the opposite side of the Earth to the centre of the earthquake, called the **epicentre**

- there is a change in direction when the P waves cross a boundary, caused by the change of speed due to a change in density.

crust

mantle

liquid outer core

solid inner core

Fig. 3.50

A transverse mechanical wave can be transmitted along the surface of a liquid, but not through its bulk.

KEY POINT

The fact that S waves form a shadow region on the opposite side of the Earth shows that part of the core is in a liquid state.

This provides evidence for the structure of the Earth shown in **Fig. 3.50**:

- a thin **outer crust** (**lithosphere**) whose thickness varies between 10 km (under oceans) and 65 km (under mountains)

- a **mantle** that behaves like a solid but allows very slow convection currents to transfer energy from the centre to the surface

- a metallic **core**, consisting mainly of nickel and iron

- the outer core is liquid but the inner core, although hotter, is solid due to the intense pressure on it.

Movement in the Earth's crust

Energy is being released in the core of the Earth due to **radioactive decay**. This results in a net energy flow from the core to the surface. The energy is carried through the mantle by slow-moving **convection currents**. These are thought to be responsible for the movement in the Earth's crust that has caused the formation of continents from a single land mass.

The mantle is usually considered to be solid, but there is a slow movement of hot rock through it.

The diagram, **Fig. 3.51**, shows how the shapes of the continents fit together, giving evidence that they could once have been joined.

Fig. 3.51

The Earth's crust consists of a number of sections called **tectonic plates**. These are moving at the rate of a few centimetres each year. The plates can:

- slide past each other
- move towards each other
- move away from each other.

When plates slide past each other, it is not a smooth movement. **Resistive forces** between the plates prevent motion until the internal forces due to **compression** and **stretching** of the material is the greater force. This results in a sudden break in the ground as the plates move with a "jerk". This movement has caused the San Andreas Fault in California.

When plates move towards each other an ocean floor plate is forced under a continental plate in a process called **subduction**. When this occurs:

- the **ocean floor plate** partially melts due to the high temperatures in the **magma**

- **metamorphic rocks** form due to the **recrystallisation** of sedimentary rocks subjected to increased temperature and pressure

- mountains can be formed due to **folds** in the continental plate; this is how the Andes mountains in South America were formed

- these mountains can be **volcanic** due to **magma** rising through faults in the crust, resulting in the formation of fine-grained rocks due to rapid cooling of the magma.

Cracks under the sea floor are due to plates moving away from each other. This is a **constructive plate margin**, as it results in the formation of new rock:

- material from the **magma** flows out of the cracks and solidifies, producing new rock

- the new rocks formed are rich in iron and become **magnetised** by the Earth's magnetic field as they solidify

- the rocks in the ocean contain a record of the direction of the Earth's magnetic field; this record shows that it **reverses** once every few thousand years.

The diagram, **Fig. 3.52**, shows **subduction** at South America and the formation of a **magnetic record** at a mid-ocean ridge.

Fig. 3.52

PROGRESS CHECK

1. State two differences between P waves and S waves.
2. What evidence is there that part of the Earth's core is liquid?
3. What does the magnetic record on the sea floor show?

1. P waves are longitudinal, S waves are transverse. P waves travel faster than S waves; 2. S waves do not travel through the Earth's core; 3. That the Earth's magnetic field reverses.

Sample IGCSE questions

1. Loudspeakers reproduce sounds which are transmitted through the air to the ears.

(a) Describe how sound travels through the air. [2]

> It travels as a longitudinal wave ✓ in which the air particles vibrate parallel to the direction of wave travel ✓.

The answer here states how the sound travels (as a longitudinal wave) and then gives a description of the motion that forms the wave to gain 2 marks.

(b) A high-pitched note has a frequency of 3500 Hz.
This travels through the air at a speed of 330 m/s.
Calculate the wavelength of the wave. [3]

> $\lambda = v/f$ or wavelength = speed ÷ frequency ✓
> $= 330 \text{ m/s} ÷ 3500 \text{ Hz}$ ✓
> $= 0.094 \text{ m}$ ✓

*The first mark is for recall and transposition of the wave equation.
Full marks are always awarded for a correct numerical answer with the correct unit. Only two marks would be awarded here if the unit was missing or wrong.*

(c) (i) Sound is diffracted as it leaves a loudspeaker.
Explain what this means. [2]

> The sound spreads out ✓ as it passes through the opening at the front of the loudspeaker ✓.

(ii) A hi-fi system has three separate loudspeakers.
Their diameters are 10 cm, 20 cm and 30 cm.
Explain which one is best for reproducing the note in (b). [4]

> The 10 cm loudspeaker is best ✓. The loudspeaker diameter is comparable to the wavelength ✓ so the sound will spread out as it leaves the loudspeaker ✓. The other loudspeakers have diameters greater than the wavelength, so the sound would not spread sufficiently to be heard everywhere in the room ✓.

(d) The diagram shows how light passes through an opening of width 10 cm. Explain why light is not diffracted in the same way as sound at a similar size opening. [3]

When answering questions about diffraction, it is important to stress that the amount of spreading depends on the size of the gap compared to the wavelength.

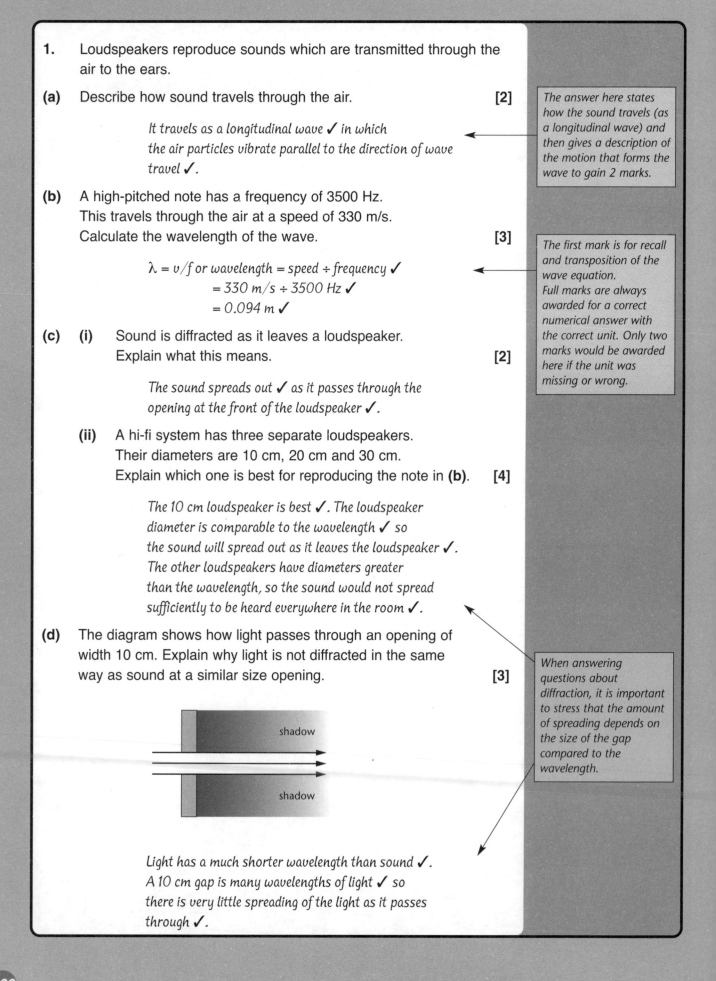

> Light has a much shorter wavelength than sound ✓.
> A 10 cm gap is many wavelengths of light ✓ so there is very little spreading of the light as it passes through ✓.

Sample IGCSE questions

2. (a) A telephone conversation can be sent either as an analogue signal along a wire or as a digital signal along an optical fibre.

Describe the difference between an analogue and a digital signal. **[2]**

> *An analogue signal varies continuously ✓.*
> *A digital signal can only have certain values, usually 0 or 1 ✓.*

Always be explicit when answering questions in tests and examinations. It is better to give too much information than too little, since extra information is not penalised.

(b) Digital signals can be *regenerated*. Explain what this means. **[3]**

> *The signal is amplified ✓ and noise is removed ✓ to return the signal to its original condition ✓.*

(c) Suggest two advantages of recording sound as a digital signal rather than an analogue signal. **[2]**

> *Noise can be removed from a recording, so there is less distortion ✓.*
> *More information can be stored in the same space ✓.*

The cue word "suggest" means that this is not required knowledge. You are expected to use knowledge of similar situations to answer this question.

(d) The diagram shows the earpiece of a telephone.

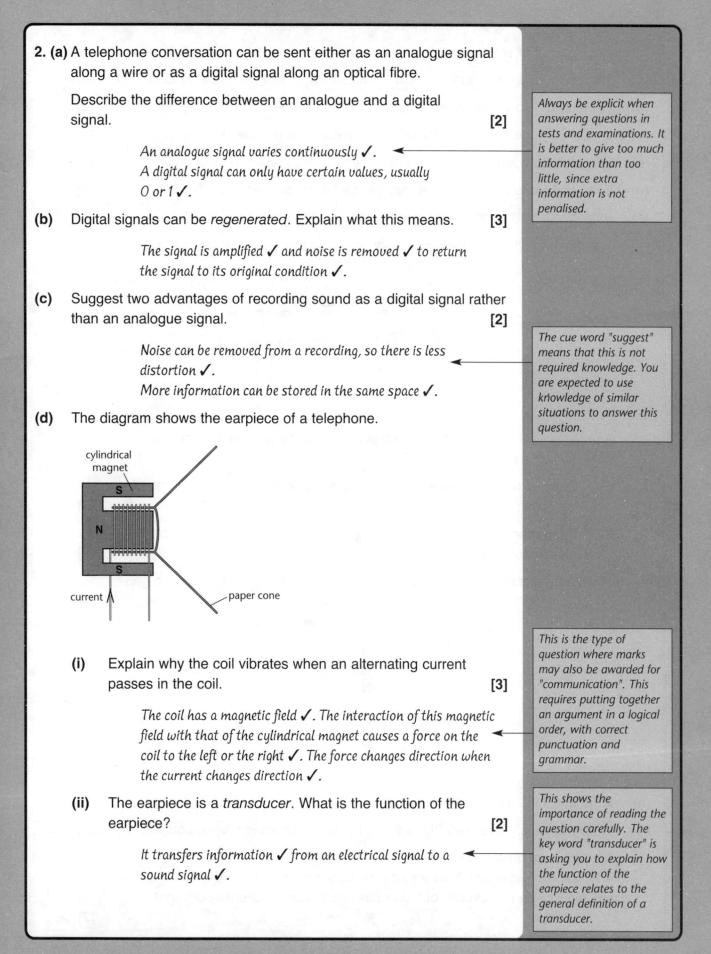

cylindrical magnet

S

N

S

current

paper cone

(i) Explain why the coil vibrates when an alternating current passes in the coil. **[3]**

> *The coil has a magnetic field ✓. The interaction of this magnetic field with that of the cylindrical magnet causes a force on the coil to the left or the right ✓. The force changes direction when the current changes direction ✓.*

This is the type of question where marks may also be awarded for "communication". This requires putting together an argument in a logical order, with correct punctuation and grammar.

(ii) The earpiece is a *transducer*. What is the function of the earpiece? **[2]**

> *It transfers information ✓ from an electrical signal to a sound signal ✓.*

This shows the importance of reading the question carefully. The key word "transducer" is asking you to explain how the function of the earpiece relates to the general definition of a transducer.

Exam practice questions

1. Here is a list of some waves:

   infra-red   sound   light   radio   ultraviolet   gamma   ultrasound

   (a) Write down two waves from the list that are transverse. **[2]**

   (b) Write down one wave from the list that can cause fluorescence. **[1]**

   (c) Which of the waves in the list is used in an electric toaster to toast bread? **[1]**

   (d) Write down one wave in the list that cannot be transmitted through a vacuum. **[1]**

   (e) Pulses of sound or ultrasound can be used to measure distances.

   The diagram shows how a house surveyor uses a sonic measuring device.

   A pulse of sound is emitted and a short time later the width of the room is displayed on a screen.

   (i) Suggest how the device uses the sound pulse to measure the width of the room. **[3]**

   (ii) Explain how the device could give unreliable results if used in a room full of people or furniture. **[2]**

2. The diagram shows how water waves spread out after passing through a gap.

   (a) Write down the name of this effect. **[1]**

   (b) What two factors determine the amount by which the wave spreads out? **[2]**

   (c) A typical sound wave has a wavelength of 1 m.
   A typical light wave has a wavelength of 5.0×10^{-7} m.
   Explain why sound spreads out after passing through a doorway but light does not. **[2]**

Exam practice questions

3. The diagram represents an object placed at a distance of 8.0 cm from a convex lens of focal length 5.0 cm.

convex lens

f

(a) Complete the ray diagram and draw in the image. **[3]**

(b) What is the distance between the lens and the image? **[1]**

(c) State two differences between the object and the image. **[2]**

(d) Which optical instrument uses a convex lens to produce this type of image? **[1]**

(e) Explain why the object is normally placed upside-down in this optical instrument. **[1]**

4. (a) What is meant by the *natural frequency* of vibration of an object? **[1]**

(b) Explain how *resonance* occurs. **[2]**

(c) A metal wire is held in tension. The tensile force remains unchanged while the length of the wire is varied. The graph shows how the natural frequency of the wire changes as its length is increased.

(i) What is the relationship between the natural frequency and the length of the wire? **[2]**

(ii) What is the natural frequency when the wire is 0.80 m long? **[1]**

(iii) What is the wavelength of the sound produced at this frequency? **[1]**

(iv) Calculate the speed of the wave along the wire. **[3]**

Exam practice questions

5. The diagram shows how an interference pattern can be observed when two beams of light overlap.

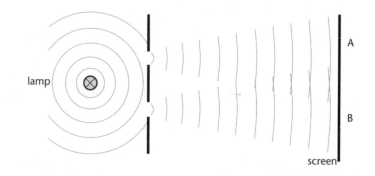

(a) Name the effect that occurs when the light passes through the narrow gaps. **[1]**

(b) Explain what is seen on the screen at A. **[2]**

(c) Explain what is seen on the screen at B. **[3]**

(d) What is meant by *path difference*? **[1]**

(e) When two wave sources are emitting waves in step, what is the path difference at points where constructive interference takes place? **[1]**

6. Magnetic tape can be used to store analogue or digital signals.

(a) Which type of signal is stored on a compact disc? **[1]**

(b) Which type of signal is supplied to a loudspeaker? **[1]**

(c) The diagram shows a tape head being used to record a signal on a tape.

(i) In what form is the signal stored on the tape? **[1]**

(ii) Describe the changes that take place as the tape passes over the narrow gap. **[2]**

(iii) A similar tape head is used to play back the recorded signal.

Explain how it does this. **[3]**

Exam practice questions

7. The diagram shows how a dish aerial is used to focus radio waves before they are transmitted to a satellite.

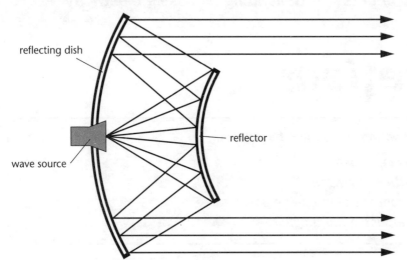

(a) Explain how focussing the beam reduces attenuation. **[2]**

(b) Radio waves can travel as ground waves, sky waves or space waves.

Explain which type of wave is used for satellite transmissions. **[2]**

(c) The satellite is in a geostationary orbit.

    (i) What are the features of a geostationary orbit? **[2]**

    (ii) What is the advantage of using a geostationary orbit for a telecommunications satellite? **[1]**

(d) The waves are received by an active satellite. What does an active satellite do? **[3]**

(e) The satellite transmission uses digital signals.

Explain why these are preferred to analogue signals. **[3]**

(f) Suggest three reasons why very short wavelength radio waves are used for the satellite transmission. **[3]**

4 Electricity and magnetism

The following topics are covered in this section:

- Magnetism ● Electric charge ● Electric circuits
- Using mains electricity ● Circuit components and digital electronics
- Electromagnetic effects ● Generating and using electricity

4.1 Magnetism

LEARNING SUMMARY

After studying this section you should be able to:

- explain induced magnetism
- recall the properties of magnets
- understand the concept of magnetic poles
- draw field lines around a bar magnet.

The phenomenon of attracting magnetic substances like iron, nickel, cobalt etc. is called magnetism.

There are two types of magnet: (i) natural and (ii) artificial.

When a magnet is brought near an unmagnetised bar of iron or steel, the bar takes on the properties of a magnet. See for yourself how a magnet can support a chain of paperclips, as magnetism is induced in each clip allowing it to attract the next one.

An unmagnetised piece of iron can be made into a permanent magnet by stroking it, in the same direction many times, with a magnet. However, the most efficient way of making a magnet is to pass an electric current through the metal to be magentised.

Natural magnets were used by early civilisations as direction finders and as aids to navigation. Now, artificial magnets can be made. A good example is steel, which has the useful property of being only temporarily magnetised when we pass an electric current through it. This makes it very useful in designing electric circuits. One easily appreciated example of this is an electromagnet crane (see picture).

Often used in the car-breaking industry, a large steel core–wound over with electric cable, is brought over a piece of metal or car-frame. When a current is turned on, the steel core becomes powerfully magnetised, attracting the metal to it and lifting it. When the crane has positioned the metal in a designated place, the current is switched off and the metal is no longer attracted to the core.

Properties of magnets

Attractive properties

A magnet always attracts magnetic substances towards itself and the force of attraction is maximum at the poles.

Directive properties

When a magnet is suspended freely by a thread, it comes to rest along a direction with one of its ends pointing towards magnetic north pole and the other towards magnetic south pole.

A magnetic pole is a point near the end of the magnet, where magnetism is concentrated. Poles always exist in pairs. To distinguish them, they are commonly known as north pole and a south pole.

 KEY POINT A single magnetic pole cannot exist.

Law of magnetic poles

Like poles repel each other and unlike poles attract each other.

To try and explain magnetism clearly, we introduce the concept of the magnetic field. Just like the gravitational field of the Earth weakens as one moves away from the Earth, so does a magnetic field weaken with increasing distance from the magnet.

If you sprinkle iron filings on a sheet of paper placed over a magnet, the filings will align themselves in a pattern which shows the effect of the magnetic field.

To help us visualise a magnetic field, we introduce the concept of **magnetic field lines** which are continuous loops that leave the north pole of a magnet and enter the south pole, continuing up inside the magnet back to the north pole. If the field is very strong, the density of lines is high i.e. there are many of them clustered together. But field lines never cross each other, they always run parallel.

Field lines can be plotted by means of a small plotting compass, laying it head to tail, marking the tip of the pointer as one proceeds from the north pole to the south pole, **Fig 4.1.**

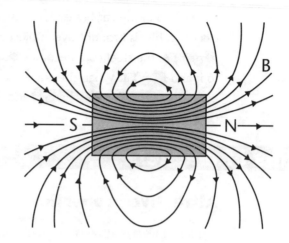

Fig. 4.1

4.2 Electric charge

After studying this section you should be able to:

● *describe how an insulating material can be charged*
● *explain the forces between charged objects*
● *recall some everyday uses and hazards of static charge.*

Creating static charge

You learned that electric **current** is moving **charge**. A concentration of electric charge that is not moving is called **static charge** or **electrostatic**. A build-up of static charge causes effects that can be hazardous or useful.

Electrons can move freely through a **conductor** but no charged particles are able to move through an **insulator**. An insulator is easily charged by rubbing:

● when an insulator is rubbed with a cloth, the friction forces between the insulator and the cloth cause electrons to be transferred from one to the other

● the object that gains electrons becomes **negatively** charged

● the object that loses electrons becomes **positively** charged.

> The body acts as a good conductor of small amounts of charge.

The diagram, **Fig. 4.2**, shows the imbalance of charge created when a polythene rod is rubbed with a duster. If a conductor such as a metal rod is rubbed while being held in the hand, any unbalanced charge is neutralised by electrons passing through the body between the rod and the earth.

Fig. 4.2

A conductor needs to be well-insulated from the Earth for it to become charged.

> The forces between two charged objects are equal in size and act in opposite directions.

If two balloons are rubbed with a cloth and then placed side-by-side, they push away from each other. Objects with opposite charges, such as one of the charged balloons and the cloth, are pulled towards each other. This shows that:

> **KEY POINT**
>
> Objects with similar charges (both negative or both positive) repel each other and those with opposite charges (negative and positive) attract each other.

The existence of repulsive forces between similar charges explains why you experience a shock if you touch an earthed conductor such as the screw on a light switch after walking across a synthetic carpet:

- synthetic carpets, for example those made out of nylon, are good insulators
- charge builds up on the body when walking across the carpet due to the friction forces between shoes and the carpet
- the similar charges repel each other but they cannot leave the body through the carpet which is a very good electrical insulator

> The direction of electron movement depends on whether the body is positively or negatively charged.

- when the body is placed in electrical contact with the earth, electrons move between the body and the earth to discharge the body, creating a current which causes a shock.

Using static charge

Electrostatic forces are used in photocopiers, inkjet printers and to paint the metal panels used in cars and washing machines.

In a **photocopier:**

- a rubber belt is coated with a material that is a conductor only when illuminated
- the belt is given a positive charge
- an image of the sheet of paper being copied is projected onto the belt, causing the illuminated parts to discharge
- the belt is sprayed with a black powder that is attracted to the charged areas

> The paper needs to be charged to attract the powder from the belt.

- the powder is transferred to a charged sheet of paper which is then heated so that the powder sticks to it.

Fig. 4.3

sheet being copied

charged belt

The diagram, **Fig. 4.3**, shows the discharge of the belt when an image is projected onto it.

In an **inkjet printer:**

- ink drops become charged as they pass through a small hole in a nozzle
- the drops pass between two parallel plates; a voltage applied to the plates deflects the drops
- the deflection of the drops can be increased by increasing the voltage and reversed in direction by reversing the voltage.

paper

negatively-charged ink drops

Fig. 4.4

If the paint is charged negatively, it repels electrons to earth, leaving the panel positively charged.

The deflection of charged ink drops is shown in the diagram, **Fig. 4.4.**

Painting metal panels uses **electrostatic induction**:

● the metal panel is connected to earth

● charged paint powder is sprayed onto the panel

● the positive charges on the paint attract electrons from the earth onto the panel

● the paint powder is attracted to the oppositely-charged panel, ensuring even coverage and little waste

● the panel is then baked in an oven to harden the paint.

Painting by electrostatic induction is shown in the diagram, **Fig. 4.5.**

Fig. 4.5

electrons flow from earth to give the panel a negative charge

Dangers of static charge

Electrostatic charge is dangerous when it causes **lightning** and **sparks** that can ignite fuel.

Charge separation results in the metal frame of an aircraft gaining an opposite charge to the fuel.

● When an aircraft is being refuelled with kerosene (paraffin) and when a car is being refuelled with petrol, **friction forces** cause **charge separation**.

● This could result in a build-up of static charge on the metal frame of the aircraft or metal sleeve of the car refuelling pipe.

● If the voltage became high enough to cause a **spark** to Earth, it could ignite the fuel.

● To prevent this, the framework of an aircraft is connected to **Earth** before refuelling and the pipe leading to the petrol tank in a car is connected to the body of the car so that the charge can spread out, preventing the build-up of charge in a small area.

---

1. Explain why a balloon is attracted to a cloth that it has been rubbed with.
2. Why does a person become charged when walking on a nylon carpet but not when walking on a woollen carpet?
3. How does a metal panel that is connected to earth become charged by electrostatic induction?

PROGRESS CHECK

1. The cloth and balloon have opposite charges; 2. Nylon is a better insulator than wool; charge from the person can pass through the woollen carpet to earth; 3. By movement of electrons between the panel and earth.

The field lines radiate outwards from a positive charge and enter a negative charge.

Just like a magnet has a magnetic field, that is represented by field lines and which becomes weaker as the distance from the magnet increases, so does an electric charge have an electric field associated with it.

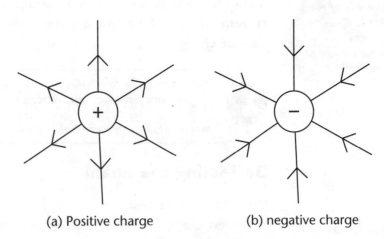

(a) Positive charge   (b) negative charge

Fig. 4.6

Electron beams

Producing a beam

The electrons in a metal wire are moving randomly at high speed. They are held within the confines of the wire by the attractive forces between them and the positive ions. Heating the wire gives the electrons more

> The electrons need energy to do work against the attractive forces.

energy and they may gain enough to leave the wire in a process known as **thermionic emission**.

An **electron gun** uses thermionic emission to produce an **electron beam**. The diagram, **Fig. 4.7**, shows the structure of an electron gun.

In an electron gun:

- the low voltage supply can be either a.c. or d.c., its job is to heat the filament that forms the **cathode**

> The electrons are repelled from the negative cathode and attracted towards the positive anode.

Fig. 4.7

- the high voltage between the **anode** and the cathode accelerates the electrons, pulling them away from the cathode, and focuses the beam
- the speed of the electrons in the beam depends on the voltage of the high voltage supply.

As the electrons leave the anode, their kinetic energy depends only on the voltage they have been accelerated through. The relationship between the kinetic energy of an electron and the accelerating voltage follows from the relationship: Work done (W) = Charge (Q) x Voltage (V)

And so energy of an electron = eV

Since energy is the ability to do work and the work done by the accelerating voltage equals the kinetic energy of the electron

> **KEY POINT**
>
> kinetic energy = electronic charge × accelerating voltage
> $$E_k = \tfrac{1}{2}mv^2 = e \times V$$

A beam of electrons consists of moving charge and so it is an **electric current**. The size of the current at the anode can be calculated as the **rate of flow of charge**, $I = Q/t$.

> **KEY POINT**
>
> The current due to a beam of electrons is given by the relationship:
> current = number of electrons per second × electronic charge
> $$I = n \times e$$
> where n is the number of electrons that pass a point in one second.

Deflecting the beam

Electron beams are used:

- in televisions, oscilloscopes and computer monitors, where the beam causes fluorescence when it strikes a coated screen
- to produce X-rays – in an X-ray tube some of the energy from the beam is emitted as X-rays when the beam strikes a tungsten target.

A beam of electrons can be deflected by **magnetic fields** and **electric fields**. Televisions and computer monitors use magnetic fields, but in an oscilloscope the beam is deflected by an electric field between a pair of parallel plates. The diagram, **Fig. 4.8**, shows the deflection of a beam of electrons by an electric field.

Fig. 4.8

Inkjet printers use a similar method for directing charged oil drops to the desired place on the paper.

In a television, a beam of electrons is directed at a screen that has three fluorescent coatings. These coatings emit light of the primary colours when the electrons strike them. The beam is made to move in a line across the screen in a process known as **scanning**. To produce a "picture", the electron beam scans the full screen with 625 lines and it completes 25 full-screen scans each second. The brightness of each colour at any point is controlled by changing the intensity of the part of the beam that strikes each of the three fluorescent layers. This is determined by the **modulation** of the signal.

Using an oscilloscope

There are two sets of deflection plates in an oscilloscope:
- the X-plates deflect the beam horizontally
- the Y-plates deflect the beam vertically.

The X-plates are used to move the beam across the screen at a constant speed. The speed is set by the time base control. When the **time base** is switched off all that is seen on the screen is a dot. Switching on the time base enables the oscilloscope to plot a graph of **voltage** against **time**. The voltage is connected to the Y-plates and causes the electron beam to move up and down, according to whether the voltage is positive or negative. The **y-sensitivity** controls the amount of deflection for each volt of the input voltage.

> When the time base is switched on, the dot appears as a line due to its rapid movement across the screen.

The diagram, **Fig. 4.9**, shows an oscilloscope trace when an alternating voltage is connected to the input, with the time base switched on.

This shows a voltage that varies in value between +6 V and –6 V.

The frequency of an alternating voltage can be calculated by measuring the time for one oscillation. In this case one oscillation corresponds to a horizontal distance of 8 cm on the screen. The time period, $T = 8 \times 0.01 = 0.08$ s. It follows that the frequency, $f = 1/T = 12.5$ Hz.

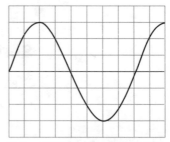

y-sensitivity = 2 V/cm   time-base = 0.01 s/cm

Fig. 4.9

PROGRESS CHECK

1. Explain how the current in an electron beam is affected by reducing the anode voltage.
2. Calculate the kinetic energy of an electron that has been accelerated through a voltage of 250 V. The electronic charge, $e = 1.6 \times 10^{-19}$ C.
3. Explain how you can tell whether a voltage is alternating or direct from an oscilloscope trace.

1. The current is reduced since fewer electrons pass through the anode each second.
2. 4.0×10^{-17} J.   3. With an alternating voltage, the trace crosses the central line. With a direct voltage, it stays on one side of this line.

4.3 Electric circuits

LEARNING SUMMARY

After studying this section you should be able to:

● *describe how the current in common circuit components varies with the applied voltage*
● *recall and use the relationship between resistance, current and voltage*
● *explain how electric current is due to charge flow.*

Measuring current and voltage

Current in a circuit is due to a flow of charge. Movement of charge is caused by the force from a voltage supply such as a battery or mains supply acting on charged particles that are free to move.

Electric current:

- is measured in **amperes** using an **ammeter** placed in **series** with the current being measured
- is not **used up** by the **components** in a circuit
- **transfers energy** from the **voltage source** to the **components**

> The term potential difference is an alternative for voltage.

Voltage:

- is measured in **volts** using a **voltmeter** placed in **parallel** with the voltage source or component
- is a measure of the **energy transfer** from a source or to a component.

Series and parallel

The diagram, **Fig. 4.10**, shows the symbols used to represent some circuit components.

> The symbols for a diode and a light-dependent resistor sometimes have circles around them – either symbol is acceptable.

cell   battery   ammeter   voltmeter   fixed resistor   variable resistor

fuse   switch   diode   light-dependent resistor   thermistor

Fig. 4.10

A **series circuit** has only one **current path**.

> The black dot in the circuit diagram shows a junction, where conductors are joined together. Its use is not essential, but helps to make the meaning clear.

The diagram, **Fig. 4.11**, shows **two lamps** connected in **series** to a **battery**, together with an **ammeter** that measures the **current** in the circuit and a **voltmeter** to measure the voltage **across** one of the lamps.

Fig. 4.11

If the voltage provided by the battery is V, then voltage across individual resistance (of lamps) will be V_1 and V_2 such that $V = V_1 + V_2$.

> **KEY POINT**
>
> In a series circuit:
> - the current is the same at all points in the current path
> - the sum of the voltages across the individual components is equal to the voltage of the power supply.

A **parallel circuit** has **more** than one **current path**.

> The circuit diagram shows two current paths.

The diagram, **Fig. 4.12**, shows two lamps connected in **parallel** to a **battery**, together with an ammeter that measures the total current in both lamps.

Here, the current, I, that passes through the battery follows **two** paths, i.e. through lamp L_1 and L_2 respectively. Thus the values of the current in these two lamps will be different i.e. I_1 and I_2 respectively, depending upon the resistance of the lamps. Most of the current goes through the path of least resistance.

Fig. 4.12

$$I = I_1 + I_2$$

> All mains circuits used at home and at work are parallel circuits. They allow appliances to be switched individually without affecting other appliances.

Thus, the current has more paths to travel and hence it encounters less resistance than if it only had one path. Hence, the compined resistance of parallel resistors is always less than any individual resistor in the combination. The voltage V, supplied by the battery, appears across each of the two lamps in parallel.

> **KEY POINT**
> In a parallel circuit:
> • all components in parallel have the same voltage across them
> • the current splits and rejoins at the junctions
> • the total current passing into a junction is equal to the current passing out of the junction.

Current and resistance

The current in a circuit or a component depends on the voltage and also the resistance. Resistance is a measure of the opposition to electric current. The higher the resistance of a component, the less current that passes in it for a given voltage.

> A variable resistor changes the current in a circuit by changing the resistance.

> **KEY POINT**
> The relationship between voltage, *V*, current, *I*, and resistance, *R* is:
> voltage = current × resistance
> In symbol form this can be written as $V = I \times R$ or $I = V/R$
> or $R = V/I$.
> The unit of resistance is the ohm (Ω).

When components are connected in series the **total** resistance is equal to the **sum** of the resistances of the individual components.

Circuit components

Common circuit components include resistors, lamps and diodes:
● the resistance of a **resistor** such as a **metal wire** does not change provided that there is no significant change in its temperature; a graph of **current** against **voltage** shows that the current is proportional to the voltage
● the wire in a **filament lamp** becomes hotter as the current in the filament increases, causing an increase in its resistance

> The direction of the current in a circuit is always shown as being from positive to negative.

● a **diode** only allows current to pass in one direction (shown by the direction of the arrow on its symbol).

The diagram, **Fig. 4.13**, shows the variation of current with applied voltage for these components.

Fig. 4.13

The resistance of some circuit components depends on their surroundings; these components are often found in electronic circuits used for switching and maintaining constant environmental conditions in, for example, greenhouses and incubators:

- the resistance of a **light-dependent resistor (LDR)** decreases with increasing light level
- the resistance of a **thermistor** decreases with increasing temperature.

The graphs in **Fig. 4.14** show the variation of resistance with environmental conditions for these components.

Fig. 4.14

A current model

Free electrons give metals the properties of being good electrical and thermal conductors.

One difference between a metallic solid such as copper and a non-metallic solid, such as salt, is that the metal contains charged particles that are moving within the body of the metal. The structure of a metal is one of fixed positive ions surrounded by a "sea" of negatively charged **free electrons**. The free electrons are said to be "free" because they have enough energy to escape from the attraction of the nucleus and move with a random motion, changing speed and direction whenever they collide with a metal ion.

Applying a voltage to the metal causes the free electrons to "drift" slowly in the direction from negative to positive, so there is an overall movement of **charge** in this direction. It is this flow of charge that forms the electric current.

KEY POINT

The relationship between charge flow and current is:
charge = current × time
$Q = I \times t$
where the charge, Q, is measured in coulombs (C) when the current is in amps and the time is in seconds.

Fluorescent lamps and street lights are examples of gases conducting electricity.

In a conducting gas or molten or dissolved electrolyte, the charge flow is due to the movement of both positively and negatively charged particles, moving in opposite directions.

Fig. 4.15

Understanding voltage

As charged particles move around a circuit, they gain energy from the battery or power supply and lose it to the components. The voltage across a power supply represents the energy transferred **to** each coulomb of charge from the power supply, and the voltage across a component represents the energy transfer **from** each coulomb of charge to the component.

> Provided that the connecting wires have a very low resistance, very little energy is transferred to them and they stay cool.

> **KEY POINT**
>
> The relationship between voltage, *V*, charge, *Q*, and the energy transferred, *E*, to or from the charge is:
>
> voltage = energy transfer ÷ charge passed
>
> *V = E/Q*
>
> This relationship also shows the equivalence of the units:
>
> 1 volt = 1 joule per coulomb

> **PROGRESS CHECK**
>
> 1. Which circuit component has a resistance that depends on the amount of light falling on it?
> 2. A current of 2.5 A passes in a wire when the voltage across it is 12.5 V. Calculate the resistance of the wire.
> 3. Name the charged particles that carry the current in a metal wire.
>
> 1. A light-dependent resistor (LDR);   2. 5.0 Ω;   3. Electrons.

Resistance and the ohm

If the current through a conductor is I when the voltage across it is V, its resistance R is defined by

$$R = \frac{V}{I}.$$

The unit of resistance is the **ohm**, which is defined as the resistance of a conductor through which a current of 1 ampere flows when a voltage of 1 volt is applied across it.

Resistors in series

The resistors in **Fig 4.16** are in series. The same current I flows through each as discussed earlier and total voltage V across all three equals the sum of separate voltages across them, i.e. $V = V_1 + V_2 + V_3$.

Fig 4.16

But $V_1 + IR_1$, $V_2 = IR_2$, $V_3 = IR_3$.

Also, if R is the combined resistance, $V = IR$,

So $IR = IR_2 + IR_2 + IR_3$ which leads to the formula for resistors in series:

$$R = R_1 + R_2 + R_3.$$

> **KEY POINT**
> In a series circuit, the total resistance equals the sum of the individual resistances.

Resistors in parallel

The resistors in **Fig 4.19** are in parallel. The voltage across the ends of each of them is the same and total current I equals the sum of the currents in the separate branches, i.e.

$$I = I_1 + I_2 + I_3.$$

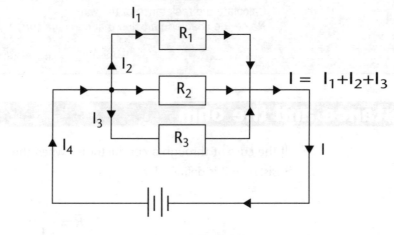

Fig 4.17

But $I_1 = \dfrac{V}{R_1}$, $I_2 = \dfrac{V}{R_2}$, $I_3 = \dfrac{V}{R_3}$.

Also if the R is the combined resistance

$V = IR,$ $I = \dfrac{V}{R}$.

So $\dfrac{V}{R} = \dfrac{V}{R_1} + \dfrac{V}{R_2} + \dfrac{V}{R_3}$ which gives the result:

$$\frac{1}{R} = \frac{1}{R_1} + \frac{1}{R_2} + \frac{1}{R_3} .$$

> **KEY POINT**
> In a parallel circuit, the reciprocal of total resistance is the sum of reciprocals of individual resistances.

Summary

Resistors in series and in parallel

You should recall that:

- in a series circuit the sum of the voltages across the individual components is equal to the supply voltage
- components connected in parallel have the same voltage across them
- the resistance of a number of resistors in series is equal to the sum of their individual resistances.

> **KEY POINT**
>
> The last point can be written as
> $$R = R_1 + R_2 + R_3$$
> where R is the total resistance of resistors R_1, R_2 and R_3 connected in series.

The diagram, **Fig. 4.18**, shows resistors connected in series and in parallel.

> A series circuit has only one current path; a parallel circuit has more than one current path.

series

parallel

Fig. 4.18

Whereas adding resistors in series always increases the resistance, adding them in parallel has the opposite effect. This is because additional resistors in parallel open up more current paths. This results in an increase in current without affecting the voltage.

> The effective resistance of a number of resistors in parallel is always less than the smallest value resistor present.

> **KEY POINT**
>
> The effective resistance of a number of resistors connected in parallel is calculated using the relationship:
> $$\frac{1}{R} = \frac{1}{R_1} + \frac{1}{R_2} + \frac{1}{R_3}$$

When using this relationship:

- calculate the value of $1/R_n$ for each resistor
- add these together
- calculate the reciprocal of this sum.

> A common error is to omit the final step.

Series and parallel combinations

The diagram, **Fig. 4.19**, shows a combination of both series and parallel resistors.

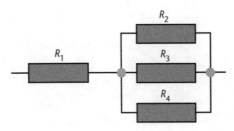

Fig. 4.19

To work out the total resistance, the combination is treated as being effectively two resistors in series:

● first calculate the effective resistance of the three resistors in parallel
● add this to the resistance of R_1.

A complete circuit

The diagram, **Fig. 4.20**, shows a combination similar to that shown above connected to a power supply to form a complete circuit.

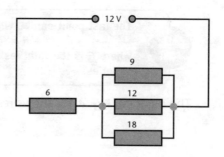

Fig. 4.20

Take care not to use the value of the circuit voltage when calculating the current in an individual component.

The current in, and voltage across, each resistor can be calculated by applying $V = I \times R$ to the complete circuit and to individual components in the following order:

● calculate the total resistance of the circuit as explained above
● use this to calculate the current that passes in the power supply and the single series resistor
● calculate the voltage across this resistor – the voltage across the parallel combination can now be worked out easily
● when the voltage across the parallel combination is known, the current in each of the parallel resistors can be calculated.

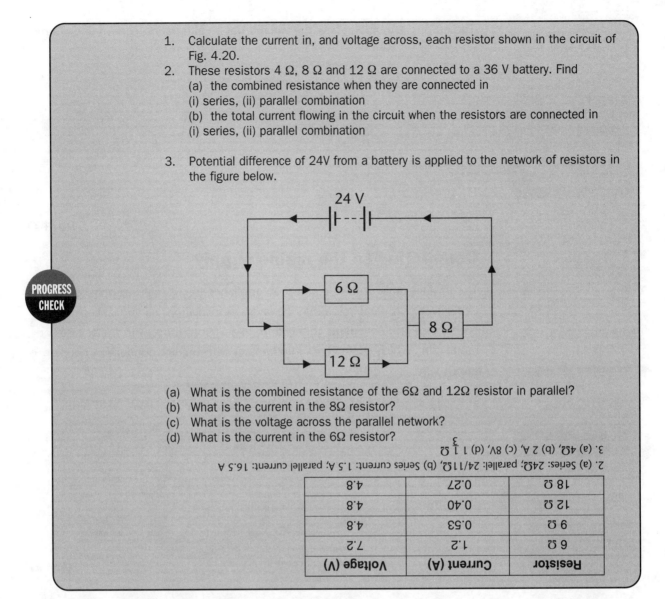

1. Calculate the current in, and voltage across, each resistor shown in the circuit of Fig. 4.20.
2. These resistors 4 Ω, 8 Ω and 12 Ω are connected to a 36 V battery. Find
   (a) the combined resistance when they are connected in
   (i) series, (ii) parallel combination
   (b) the total current flowing in the circuit when the resistors are connected in
   (i) series, (ii) parallel combination

3. Potential difference of 24V from a battery is applied to the network of resistors in the figure below.

24 V

6 Ω

8 Ω

12 Ω

(a) What is the combined resistance of the 6Ω and 12Ω resistor in parallel?
(b) What is the current in the 8Ω resistor?
(c) What is the voltage across the parallel network?
(d) What is the current in the 6Ω resistor?

3. (a) 4Ω, (b) 2 A, (c) 8V, (d) 1 $\frac{1}{3}$ Ω

2. (a) Series: 24Ω; parallel: 24/11Ω, (b) Series current: 1.5 A; parallel current: 16.5 A

Resistor	Current (A)	Voltage (V)
6 Ω	1.2	7.2
9 Ω	0.53	4.8
12 Ω	0.40	4.8
18 Ω	0.27	4.8

4.4 Using mains electricity

After studying this section you should be able to:

● **recall the difference between alternating current and direct current**
● **explain the functions of the wires that connect an appliance to the mains supply**
● **describe how electricity from the mains supply is used and measured.**

Alternating and direct current

All circuits need a source of energy. Batteries are used in appliances such as torches and portable music players. Batteries use expensive chemicals, so the mains electricity supply is used for appliances used in and around the home. The current in a battery-powered circuit is **direct current (d.c.)**, while that in a mains-powered circuit is **alternating current (a.c.)**:

● direct current passes in the same direction
● alternating current changes direction.

The graphs, **Fig. 4.21**, show the variation of an alternating current and a direct current.

A direct current does not have to maintain a constant value. The size of the current can change, but the direction is always the same.

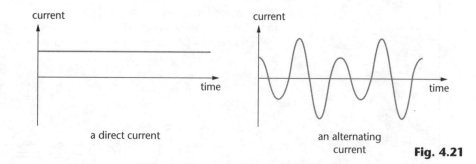

a direct current

an alternating current

Fig. 4.21

Connecting to the mains supply

The flexible cable used to connect an appliance to the mains supply contains either two or three wires within the layers of insulation. All appliances need both a **live** and a **neutral** connection. In addition, those with metal cases or metal parts which could come into contact with the live connection need an **earth** wire.

The diagram, **Fig. 4.22**, shows how a kettle is connected to a mains plug.

Each wire in the mains cable has two layers of insulation. The colour of the inner layer is different for each conductor. Notice that the cable grip should clamp the outer insulation.

earth

live

neutral

fuse

live and neutral to heating element

cable grip covers outer insulation

earth to metal element casing

Fig. 4.22

When the kettle is switched on:

- energy is supplied through the **live** conductor (brown); the voltage of this alternates between positive and negative
- the **neutral** conductor (blue) is there to form a complete current path or circuit
- the **earth** wire (green and yellow stripes) is needed for safety
- the **fuse** is also present for safety – it breaks the circuit if the current becomes abnormally high.

The current rating of the fuse should be slightly higher than the current when the appliance is operating normally. A fault such as a **short-circuit** would increase the current passing in the appliance and the wires, causing them to overheat and creating a fire hazard. The fuse prevents this by melting and breaking the circuit when the current exceeds the rating of the fuse.

A fault that causes the metal casing to become live could electrocute the user. The earth wire and fuse together guard against this. If this happens the low resistance path from live to earth causes a large current in the fuse, breaking the circuit. It is important that the fuse is placed in the live conductor so that it cuts off the voltage supply.

> If the appliance has a plastic casing, the casing cannot conduct electricity.

Some appliances are said to be **double-insulated** and do not need an earth connection. A double-insulated appliance:

- has a plastic casing with no exposed metal parts
- cannot cause electrocution since the casing cannot become live
- carries this symbol:

Circuit breakers are now used in place of **fuses** to protect the fixed cables that connect sockets to the mains supply from overheating. The advantages of using a circuit breaker instead of a fuse are:
- a circuit breaker is easily reset when a fault has been rectified
- a circuit breaker acts faster than a fuse, giving greater protection.

> A person standing inside has a much greater resistance than one outside because there is more insulation between him and the ground.

When using appliances such as lawnmowers and hedge trimmers outdoors, there is an increased risk of electrocution. This is because these appliances can readily cut through a cable and there is little insulation between a person and the earth. A **residual current circuit breaker (RCCB)** should always be used to connect the appliance to the mains supply. An RCCB detects any difference between the current in the live and neutral wires. If the current in the neutral wire is less than that in the live, which happens if current passes in a person, then it cuts off the supply voltage.

Energy transfer and cost

Electrical appliances used at home transfer energy from the mains supply to:
- heat
- light
- movement (including sound).

> Sound from a television or radio is produced by the movement of a loudspeaker cone.

The power of an appliance is the rate at which it transfers energy.

> **KEY POINT**
> Electrical power is calculated using the relationship:
> power = current × voltage
> $P = I \times V$
> Power is measured in watts (W) where 1 W = 1 J/s.

Appliances used for heating have a much higher power rating than those used for lighting or to reproduce sound. Heating can come about in a number of different ways:

> All wires have some resistance. The greater the resistance, the greater the heating when a given current passes.

- by **convection currents**, when a hot wire heats the surrounding air or water, as in a kettle
- by **infra-red radiation** given off by a wire hot enough to glow red, as in a toaster
- by absorption of **microwaves** by water molecules in food, as in a microwave cooker.

Fig. 4.23

The quantity of **energy** transferred from the mains by an appliance depends on:

- the **power** of the appliance

- the **time** for which it is switched on.

 Since P = IV, this is also equivalent to E = IVt.

> **KEY POINT**
>
> **Energy transfer from electricity is calculated using:**
> energy = power × time
> $E = P \times t$
> **The energy transfer is in J when the power is measured in W and the time is measured in s.**

The electricity meter in your home measures the energy transferred from the mains supply. If it measured the energy in joules, it would need a lot of digits as the joule is a very small unit of energy compared to the amount transferred to a typical house each day. Instead of the joule, electricity companies measure the energy supplied in **kilowatt-hours** (kW h).

> **KEY POINT**
>
> **One kilowatt-hour is the energy supplied to a 1 kW appliance when it operates for 1 hour.**

To calculate energy transfer in kW h:

- the equation *energy = power × time* is used

- the power is measured in **kilowatts (kW)**

- the time is measured in **hours (h)**.

 The term "unit" is sometimes used to refer to 1 kilowatt-hour.

The cost of each kilowatt-hour of energy from the electricity mains supply varies but it is currently about 7p. An electricity bill is calculated by multiplying the number of "units" supplied by the cost of each one.

> **KEY POINT**
>
> **The cost of energy from electricity is calculated using:**
> cost = power in kW × time in h × cost of 1 kW h
> or
> cost = number of kW h × cost of 1 kW h.

4.5 Circuit components and digital electronics

LEARNING SUMMARY

After studying this section you should be able to:

- *distinguish between n and p-type semiconductors*
- *understand the concept of holes*
- *describe the action of diode*
- *understand the working of a transistor*
- *identify different parts of an electronic system*
- *describe how logic gates are used as processors.*

In the past decades, new discoveries have led to the development of many electronic devices such as radios, mobile phones, digital storage of music etc. The list is long and new products are being devised with ever-increasing pace. All this has been possible by the use of semiconducting materials like silicon (hence 'silicon chips') and germanium.

Although substances are often divided into two categories on the basis of their conductivity – good conductors (such as metals) and bad (such as wood and plastic) conductors of electricity, a category which now rivals them in importance is that of semiconductors. As the name implies, in the natural state, semiconductors have a conducting ability less than that of conductors but greater than insulators. They are of two kinds:

- a negative semiconductor (n-type) has a surplus of electrons, and so its charge carriers are negative electrons

- a positive semiconductor (p-type) has charge carriers that are positive holes.

A hole is actually an absence of an electron. Hence, under the influence of an electric field, other electrons move in to fill the hole. The net effect of electrons moving one way is the same as a hole moving in the opposite direction.

Fig. 4.24 n–type semiconductor

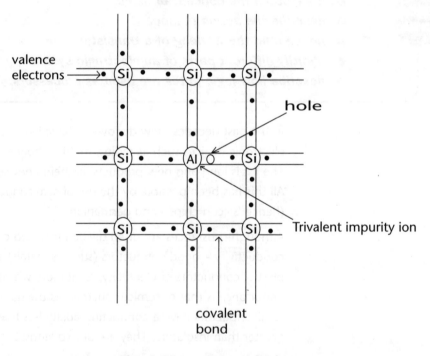

Fig. 4.25 p–type semiconductor

We can introduce more electrons or holes into a semiconductor by diffusing an element with more or less electrons into its structure, thus controlling the conducting ability. The two major uses of this technique are in **diodes** which act as a valve, or in **transistors**, which act as a switch.

A diode consists a of a p-n junction, that consists of a p-type semiconductor and an n-type semiconductor joined together. When they are joined, the holes and electrons at the junction combine to form a small neutral zone.

When placed in a circuit with the n-type side connected to a negative voltage and the p-side to a positive voltage, the diode is said to be forward biased. In forward bias configuration, a current flows as the electrons are attracted across the junction to the positive side. When the diode is connected the opposite way around, the electrons are held back from crossing the p-n junction and no current flows (a small leakage current may pass through). So the diode acts as a one-way valve, allowing the current to pass only in one direction and not the other.

(a)

(b)

Fig. 4.26

A transistor has three connections (known as the **collector**, the **base** and the **emitter**) and its structure is best described as two diodes placed back-to-back. It acts as an amplifier – in effect, it allows a tiny current to control a much larger one. A silicon chip contains many transistors, each functioning as a part of an integrated circuit.

Electronic systems

An electronic system can react to changes in environmental conditions such as pressure, illumination and temperature. There are three components to an electronic system:

- **input sensors** provide information in the form of a voltage that can have different values

- **processors** receive this information and produce an output which depends on the inputs

- **output devices** are operated by the output voltage from the processor.

> In a calculator the keypad provides the input to the processor, and the display is the output device.

Input devices include:

- **switches** – these respond to changes in pressure, magnetic fields or tilting

- **moisture detectors** – these have a resistance that depends on the amount of water present

- **light-dependent resistors (LDRs)** – these have a resistance that depends on the illumination

- **thermistors** – these have a resistance that depends on the temperature.

> In this chapter, the term thermistor refers to one with a negative temperature coefficient – its resistance decreases with increasing temperature.

Output devices include **lamps** and **light-emitting diodes (LEDs)**, but in many applications a relay is used to switch a high-current or high-voltage device such as a heater or a motor.

Logic gates

Many electronic systems use processors based on **logic gates**. The signals in such a system are **digital** as they can only have one of two values, on (high) or off (low). Most logic gates work from a potential difference (voltage) of 5 V, so an input or output voltage of 0 V is on or low and one of +5 V is on or high. The way in which the output signal depends on the inputs is shown in a **truth table**. In a truth table a 0 represents off and a 1 represents on.

> Voltage is measured relative to the negative terminal of the battery or power supply.

The diagram, **Fig. 4.26**, shows the circuit symbols for NOT, OR and AND logic gates, together with their truth tables.

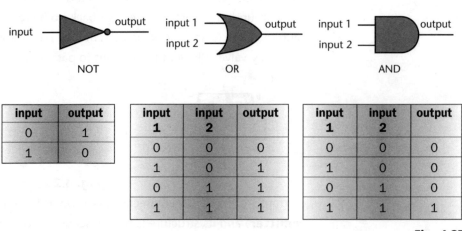

input	output
0	1
1	0

input 1	input 2	output
0	0	0
1	0	1
0	1	1
1	1	1

input 1	input 2	output
0	0	0
1	0	0
0	1	0
1	1	1

Fig. 4.27

A NOT gate is also referred to as an invertor.

The name of each gate describes how the output depends on the inputs:
- NOT – the output is 1 if the input is 0 or NOT 1; this gate inverts an input signal
- OR – the output is 1 if either input 1 OR input 2 is 1 OR both are 1
- AND – the output is 1 if input 1 AND input 2 are both 1.

Combining logic gates

Outputs from logic gates can be used as inputs to other gates. Combinations of logic gates are used to perform a variety of functions. The diagram, **Fig. 4.28**, shows how NAND and NOR gates can be made from combinations of NOT, AND and OR gates. The truth tables for these gates are also shown.

NAND means NOT AND and NOR means NOT OR.

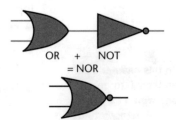

input 1	input 2	output
0	0	1
1	0	1
0	1	1
1	1	0

input 1	input 2	output
0	0	1
1	0	0
0	1	0
1	1	0

Fig. 4.28

The outputs of these gates are the inverse of those of AND and OR.

Controlling the input

Switches can be used to apply a voltage (0 V or +5 V) directly to the input of a logic gate. Other input sensors are used in **potential divider** circuits to give a voltage that varies with environmental conditions. The diagram, **Fig. 4.29**, shows a potential divider circuit that uses a fixed resistor and a variable resistor.

Depending on the values of the resistors, the output voltage of a potential divider can range from 0 V to the full supply voltage.

Fig. 4.29

In this potential divider circuit:

- the input voltage is equal to the sum of the voltages across the resistors
- the voltage is divided in the ratio of the resistances, so the larger value resistor has the greater voltage across it
- decreasing the resistance of the variable resistor causes the output voltage to rise.

KEY POINT

The relationship between the output voltage, V_{out}, the input voltage, V_{in} and the resistor values in a potential divider circuit is:

$$V_{out} = V_{in} \times \frac{R_2}{R_1 + R_2}$$

where the output voltage is that across resistor R_2.

The diagram, **Fig. 4.30**, shows potential divider circuits that use a thermistor and an LDR to provide the input signals to logic gates.

Fig. 4.30

a temperature-dependent potential divider   a light-dependent potential divider

The LDR symbol is sometimes drawn with a circle around the rectangle.

In the temperature-dependent potential divider:

- reducing the temperature increases the resistance of the thermistor
- this causes the voltage across the thermistor to rise
- the signal changes from 0 to 1 when the voltage crosses the threshold voltage
- the temperature at which this happens is determined by the value of the fixed resistor; if a variable resistor is used instead this can be adjusted easily.

The threshold voltage is the minimum voltage that the logic gate recognises as being 1. It is usually about half the supply voltage.

A similar effect occurs when the illumination of the LDR falls in the light-dependent potential divider.

With the fixed (or variable) resistors and the thermistor or LDR in the **same** positions, the output voltage can be made to fall as the temperature or illumination increases.

Using the output

An LED can be used to show whether the output of a logic gate is 0 or 1. The output voltage is too high for an LED, so a series resistor is used to limit the current. This is shown in the diagram, **Fig. 4.31**.

> The LED symbol, like the LDR symbol, is sometimes drawn with a circle around it.

0 V

Fig. 4.31

Logic gates can only supply a small current so they can only be used to drive low-power devices. Using the logic gate to switch a relay enables the switching of higher-power appliances which can operate from a separate, independent voltage supply such as the mains.

The diagram, **Fig. 4.32**, shows a logic gate being used to switch a mains-operated lamp. The NO on the relay symbol means that the switch contacts are normally open (switched off). They close, switching on the circuit, when current passes in the relay coil.

> The rectangle is the symbol for a relay coil.

NO

0 V

Fig. 4.32

A complete system

A microelectronic system is required to control a mains-operated electrical heater in a commercial greenhouse. The heater:

● should never be on during the day
● should switch on at night when the air temperature drops below a certain level.

> When deciding on the processor to use, write down the conditions under which it is to be switched on, using the logic words AND, OR etc.

Two sensors are required for this system, one to sense whether it is day or night and one to sense the temperature. The potential divider circuits shown in the diagram, **Fig. 4.33**, can be used as inputs. The logic gate should switch on power to the heater when it is cold and dark, so an AND gate is the processor to be used in such a case. As the heater works from the mains supply, the output device is a relay to switch the heater on and off.

The diagram, **Fig. 4.33**, shows an electronic system that satisfies the requirements.

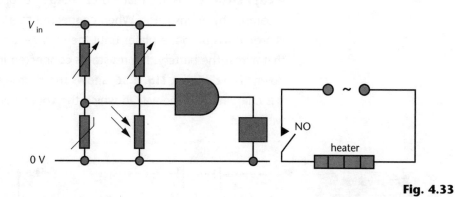

Fig. 4.33

The latch

An electronic system such as a burglar alarm has several sensors that are activated by opening a door or window or movement within a building. Once the alarm has been triggered, it should keep ringing even if the door or window is closed or the burglar stands still. A **latch** is used to maintain the output even if the input from a sensor changes from 1 to 0.

A bistable latch:

● has two stable output states, 0 and 1
● maintains this output when the input signal is removed
● can be set and reset by applying a 1 momentarily to the appropriate input.

The diagram, **Fig. 4.34**, shows a latch which uses two NOR gates.

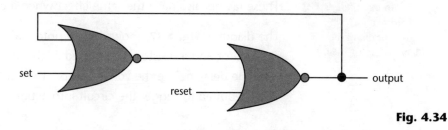

Fig. 4.34

A bistable latch can also be made from NAND gates. In the circuit shown in the next diagram, **Fig. 4.35**, the two NAND gates on the left are acting as invertors; if these are omitted, the latch is set and reset by applying a momentary 0 rather than a 1.

Fig. 4.35

The timing circuit

A **capacitor** is a device that stores charge. It consists of two metal plates separated by an insulator. When connected to a battery or power supply, charge flows onto the plates until the voltage across the capacitor is equal to that across the battery. If a resistor is connected in series the process is slowed down. The diagram, **Fig. 4.36**, shows the charge flow onto a capacitor while it is charging and the way in which the voltage across the capacitor changes with time.

> The variation of voltage with time shows an exponential rise. When a capacitor discharges, the voltage decreases exponentially. The graph has the same shape as a radioactive decay curve.

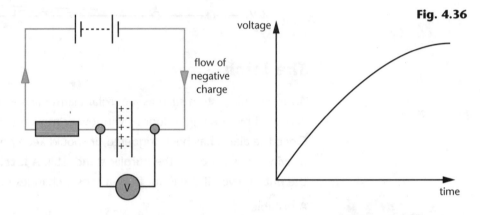

Fig. 4.36

The time taken for the voltage to reach a given value depends on:
- the resistance of the resistor; the greater the resistance, the longer it takes for the capacitor to charge
- the capacitance of the capacitor; the greater the capacitance, the longer it takes for the capacitor to charge.

> Capacitance is measured in farads (F). This is a large unit; the values of practical capacitors are usually measured in microfarads ($1\ \mu F = 1 \times 10^{-6}\ F$) or picofarads ($1\ pF = 1 \times 10^{-12}\ F$).

These factors also have the same effect when a capacitor discharges.

The diagram, **Fig. 4.37**, shows how a potential divider circuit can be made from a capacitor and a resistor. When used as the input to a switching circuit, there is a time delay before the voltage across the capacitor reaches the threshold value required to trigger the circuit into action.

> The time delay is determined by the values of the resistor and the capacitor.

Fig. 4.37

Each time the circuit is used, the capacitor needs to be discharged. This is done by closing the "reset" switch.

Amplifying the current

If a logic gate circuit is used to drive a relay which requires a higher current than the logic gate can deliver, a **transistor** is used. A small output current from the logic gate is used to switch a much larger current in the transistor, which in turn operates the relay. The diagram, **Fig. 4.38**, shows a light-dependent switch that uses a transistor to switch the current in a relay coil.

> By changing the input sensor, this circuit can be used to switch on devices when there are changes in light level, temperature or humidity.

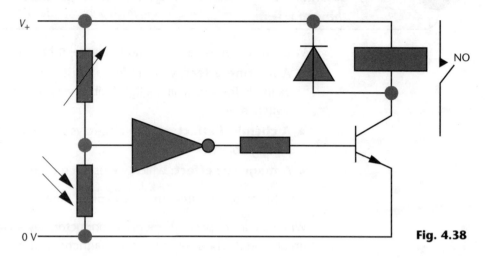

Fig. 4.38

In this circuit:

● as the illumination of the LDR is increased, its resistance decreases and the voltage across it drops (voltage across variable resistor increases).

● the **input** to the NOT gate decreases to **below** threshold (a 0) – so the **output** from the NOT gate changes to 1. This starts the transistor conducting, thus operating the relay.

● the variable resistor can be adjusted to change the light level at which the logic gate is switched on

● the diode prevents a large induced voltage when the relay switches off – this would damage the transistor.

1. In a radio, what is:
   (a) the input sensor
   (b) the output device?
2. Which logic gate should be used as the processor in a burglar alarm which has two sensors?
3. Explain why a relay is needed for a logic gate to switch a mains-powered appliance.

1a the aerial; b the loudspeaker; 2 OR; 3 The logic gate operates from a low voltage and can only supply a small current. The mains lamp needs a much higher current at a higher voltage.

4.6 Electromagnetic effects

LEARNING SUMMARY

After studying this section you should be able to:

● *enumerate different effects of current*
● *determine the direction of magnetic fields produced by current flowing through a wire.*

When a current is passed thruogh a wire, it has three effects.

● **A heating effect**: you can feel a resistor in a circuit get warm. As an example from daily life, a lightbulb is very hot to the touch when the light is switched on.

● **A chemical effect**: seen in electrolysis – the process that we use in silver plating objects.

● **A magnetic effect**: which we use in motors.

It is this magnetic effect that we explore here.

When a current passes through a conductor, say a wire, a magnetic field is set up around it. The strength of the magnetic field depends on the current passing through the wire.

KEY POINT   In order to ascertain the direction of field lines around the wire, we use the right-hand current rule convention.

In diagrams, we use a dot to symbolise the point of an arrow, and hence a current coming towards us. A cross is used to symbolise an arrowhead moving away from us.

If we imagine the wire to be held in one's right hand, such that the thumb points in the direction of the current, then the fingers curling around the wire give the direction of field lines.

Use the right-hand current rule to verify that the circles representing the magnetic field in **Fig. 4.39** are drawn in the correct direction.

Fig. 4.39

4.7 Generating and using electricity

After studying this section you should be able to:

● describe how the magnetic force on a current is used in a motor
● explain how electromagnetic induction is involved in generating and distributing electricity
● evaluate the use of different energy resources for producing electricity.

The motor effect

It has been observed that a **force** acts on an **electric current** that passes in a direction at right angles to a **magnetic field**.

The direction of this force:

● is at right angles to both the magnetic field and the current direction

● is reversed if either the current or the magnetic field direction is reversed.

Fig. 4.40

The size of the force:

● is increased by increasing the current or the strength of the magnetic field.

> The strength of the magnetic field can be increased by using stronger magnets.

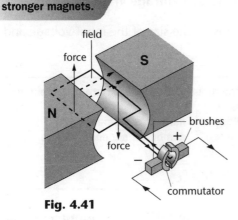

Fig. 4.41

The diagram, **Fig 4.41**, shows how this is used to produce rotation in a simple **d.c. motor**. The purpose of the **commutator** is to reverse the direction of the current as the coil passes through the vertical position. This is necessary to keep the loop turning in the same direction.

The motor effect is also used to produce the vibrational movement of a **loudspeaker** cone. A coil of wire within a magnetic field carries an **alternating current**. The force on the coil is reversed, reversing the direction of movement, whenever the current changes direction.

Bicycle **dynamos** also use a rotating magnet, see **Fig 4.42**. As a bicycle speeds up the magnet rotates faster, increasing the size of the induced voltage so the lights become brighter. It also increases the **frequency** of the induced voltage, since one cycle is generated for each revolution of the magnet.

Fig. 4.42

The transformer

> As they rely on changing magnetic fields, transformers cannot be used to change the size of a direct voltage.

A **transformer** changes the size of an **alternating voltage**. The diagram, **Fig. 4.43**, shows the construction of a transformer. It works on the principle of electromagnetic induction:

- an alternating current in the input or primary coil produces a changing magnetic field

- this changing field is concentrated in the iron core, so that it passes through the output, or secondary coil

Fig. 4.43

- the **changing magnetic field** induces a **voltage** in the secondary coil.

The size of the output voltage depends on the size of the input voltage and the ratio of the numbers of turns on the coils.

> This formula is cumbersome to use. When calculating transformer voltages, it is usually easier to work in ratios.

KEY POINT

The relationship between the input (primary) and output (secondary) voltages of a transformer is:

$$\frac{\text{primary voltage}}{\text{secondary voltage}} = \frac{\text{number of primary turns}}{\text{number of secondary turns}}$$

$$\frac{V_p}{V_s} = \frac{n_p}{n_s}$$

This means that the voltages are in the **same ratio** as the **numbers of turns** on the coils:

- a **step-up** transformer **increases** the voltage; it has more turns on the secondary than on the primary

- a **step-down** transformer **decreases** the voltage; it has fewer turns on the secondary than on the primary.

When a transformer is used to **increase a voltage**, the **current is reduced** by the same factor. Similarly, decreasing the size of a voltage results in an increased current.

Electromagnetic induction

When the **magnetic field** through a coil changes, it causes a **voltage** across the terminals of the coil. This voltage is called an induced voltage and the phenomenon is known as **electromagnetic induction**. A voltage can be induced in a conductor by:

- moving the conductor at right angles to a magnetic field
- moving a magnet inside a coil of wire
- switching a nearby electromagnet on or off.

> When answering questions about electromagnetic induction, always emphasise the change in the magnetic field.

The diagram, **Fig. 4.44**, shows a **voltage** being **induced** as a magnet is moved into a coil of wire. A **current** passes if there is a complete circuit.

The size of the induced voltage can be increased by:

- increasing the number of turns on the coil
- increasing the area of the coil
- moving the magnet faster
- increasing the strength of the magnetic field.

> Moving the coil around a stationary magnet would have a similar effect.

Fig. 4.44

The direction of the induced voltage can be reversed by:

- reversing the direction of movement
- reversing the poles of the magnet.

The generator

If a **coil** of wire is rotated within a **magnetic field**, electromagnetic induction causes a **voltage** across the connections to the coil. The diagram, **Fig. 4.45**, shows an **a.c. generator**. The slip rings and brushes enable the induced current to pass out of the coil.

> Rotating the coil causes a change in the magnetic field through it.

Generators like this are used to provide the electricity supply for motor vehicles.

Power station generators use a rotating **electromagnet** to produce a changing magnetic field. A voltage is induced in thick copper bars around the electromagnet. The electromagnet rotates at 3000 revolutions each minute to induce a voltage with a frequency of 50 Hz.

> It would be impractical to rotate the copper bars at this speed.

coil

slip rings

brushes

Fig. 4.45

Transmitting power

Whenever a current passes through a wire, the resistance of the wire causes **energy loss** due to heat. The energy losses can be minimised by having very low resistance wires. This leads to conflict between installation costs and running costs. The conflict is resolved by using transformers:

- the rate of energy transfer to heat in transmission wires is proportional to the $(\text{current})^2$

- **low currents** need to be used to minimise energy losses

- this can be achieved by transmitting power at **high voltages**.

A typical power station generator produces electricity at a voltage of 25 000 V. This is stepped up by a transformer before passing into the grid at 400 000 V. The high voltage electricity is stepped down in stages before it reaches consumers such as transport, industry and houses. This process is shown in the diagram, **Fig. 4.46**.

Fig. 4.46

Transformers provide an efficient way of changing the voltage of an alternating current. There is no equivalent way of changing the voltage of a direct current, which is why we use alternating current for mains electricity.

Energy resources

Most of the electricity generated nowadays comes from **fossil fuels** – coal, gas and oil. The amount of electricity produced from nuclear energy is significant, but decreasing. Some electricity is generated from moving water in fast-flowing rivers and tides and a small amount comes from **wind-powered** generators and **geothermal** energy.

In a **coal-fired** power station:

- energy obtained from burning coal is used to turn water into steam at high temperature and pressure

- this steam drives the **turbines** which in turn drive the **generators**

- a significant energy loss occurs when the steam is condensed back into water to be pumped back to the boiler.

In a **gas-fired** power station:

- gas is burned in a combustion chamber
- the hot exhaust gases drive the turbines directly
- the energy remaining in the exhaust gases is used to generate steam to drive a steam turbine
- this process has an overall **efficiency** of 50% – 60%, which is greater than a coal-fired power station.

Although the known reserves of coal are greater than those of gas, new gas fields are constantly being discovered. The building of gas-fired stations to replace older coal-fired ones helps to reduce **carbon dioxide** emissions.

The burning of fossil fuels causes **pollution** in the forms of **carbon dioxide** – a greenhouse gas, and **sulphur dioxide** – which contributes to **acid rain**. Sulphur dioxide can be removed from power station exhaust gases by passing them through a slurry of powdered limestone mixed with water. However, this leads to other forms of pollution in mining and transporting the limestone.

> Burning gas causes less sulphur dioxide pollution than oil, which in turn causes less than coal.

In a **nuclear** power station:

- energy released from uranium and plutonium is used to create steam and drive a steam turbine, as in a coal-fired station.
- There are **environmental hazards** due to the accidental release of radioactive material.
- The **waste** materials can be highly **radioactive** and present long-term disposal and storage problems.

> The high cost of closing down a nuclear power station adds significantly to the cost of the electricity produced.

The use of renewable energy resources such as **wind** and **moving water** has the advantage of producing no atmospheric pollution:

- fast-flowing rivers and streams are used to drive water turbines in the production of **hydro-electricity**; suitable rivers and streams are only found in Wales and Scotland
- there is a vast amount of energy available from **tides**; this does not depend on rainfall or wind and is constantly available but attempts to transfer this energy to electricity have proved costly and unreliable

> Building a barrier across a tidal estuary can cause silting-up and consequent environmental damage.

- in some parts of the country the wind blows all the time. In such places, **wind-powered** generators can be relied upon; however, wind farms have a high installation cost, they are regarded as noisy and unsightly by some people and they take up a large area to produce a relatively small amount of electricity
- underground rocks are constantly being heated by **radioactive decay**; the **geothermal energy** can be extracted by pumping water through them, but there are few sites in the UK where the rocks are hot enough to generate steam capable of driving turbines.

Capturing energy from the Sun

Radiant energy from the Sun can be used to heat water and generate electricity. The use of **solar heating** to provide domestic hot water is very common in southern Europe where the intensity of the Sun's radiation is greater than that in the UK, so it takes less time to recoup the capital outlay in lower fuel bills.

storage tank

insulation

cold water

blackened copper pipes

glass-covered box

dull black surface

Fig. 4.47

The diagram, **Fig. 4.47**, shows a solar heated water panel and hot water storage tank:

- **short-wavelength** radiation from the Sun passes through the glass panel and is absorbed by the blackened copper pipes

- energy passes through the copper pipes by **conduction** to heat the water

- **long-wavelength** infra-red radiation emitted by the water pipes does not pass through the glass panel; much of it is reabsorbed by the pipes.

Photovoltaic cells can be used to clad the walls and roofs of buildings and generate electricity. At the moment these have a very low **efficiency** (around 20%) and a very high capital cost. They are useful for low-power appliances such as calculators and to power telephone boxes in remote areas, where it is cheaper to install the solar cells than to provide a connection to the mains electricity supply.

PROGRESS CHECK

1. What two changes take place to the output of an a.c. generator when it is turned faster?
2. A transformer has an input voltage of 240 V and 1 000 turns on the primary coil. How many turns are needed on the secondary coil for the output voltage to be 12 V?
3. Explain why electricity is transmitted at high voltage and low current.

1. The voltage and frequency both increase;   2. 50;   3. To minimise energy lost as heat in the wires.

Sample IGCSE questions

1. The table shows the current that passes in three household appliances when they are connected to the 240 V mains supply.

Appliance	Current in A
grill	6.0
desk lamp	0.6
convector heater	4.5

(a) (i) Which of these has the greatest resistance? Explain how you can tell. **[2]**

The lamp ✓ as it has the least current for the same voltage ✓.

> It is important here to refer to the data in the table and the introduction to the question.

(ii) Calculate the resistance of the grill. **[3]**

$R = V/I$ or resistance = voltage ÷ current ✓
$= 240\ V ÷ 6.0\ A$ ✓
$= 40\ \Omega$ ✓

> These examples show how to set out calculations based on formulas that are either given or that you have to recall. In each case always write down the formula that you are using, then write it with the values in place and finally give the answer with the correct unit.

(iii) Calculate the power rating of the convector heater. **[3]**

$P = IV$ or power = current × voltage ✓
$= 4.5\ A × 240\ V$ ✓
$= 1080\ W$ ✓

(b) The plug fitted to the convector heater has three wires: live, neutral and earth.

(i) Which of these transfers energy from the mains supply? **[1]**

The live wire ✓.

> The allocation of two marks shows that two separate points are needed. One mark is for realising that the wires would be heated and the second mark is for giving a reason why this is not desirable.

(ii) Why is it important that the wires connecting the heater to the mains supply have a low resistance? **[2]**

So that the wires do not become hot ✓.
Which would waste energy and be a fire hazard ✓.

(iii) Explain how the earth wire, along with the fuse fitted to the plug, guard against electrocution. **[3]**

If a fault occurs so that the casing becomes live, a large current passes to earth ✓.
This melts the fuse wire ✓.
Which breaks the connection in the live wire so that the casing is no longer live ✓.

> This type of question, requiring a full explanation written in a logical order, could carry additional marks for the quality of your written communication. You will not gain these marks if the order of your answer is illogical or your use of sentences is poor.

Sample IGCSE questions

2. Two coils of wire are wound on an iron core.
    One coil is connected, through a switch, to a d.c. supply.
    The second coil is connected to a sensitive ammeter.

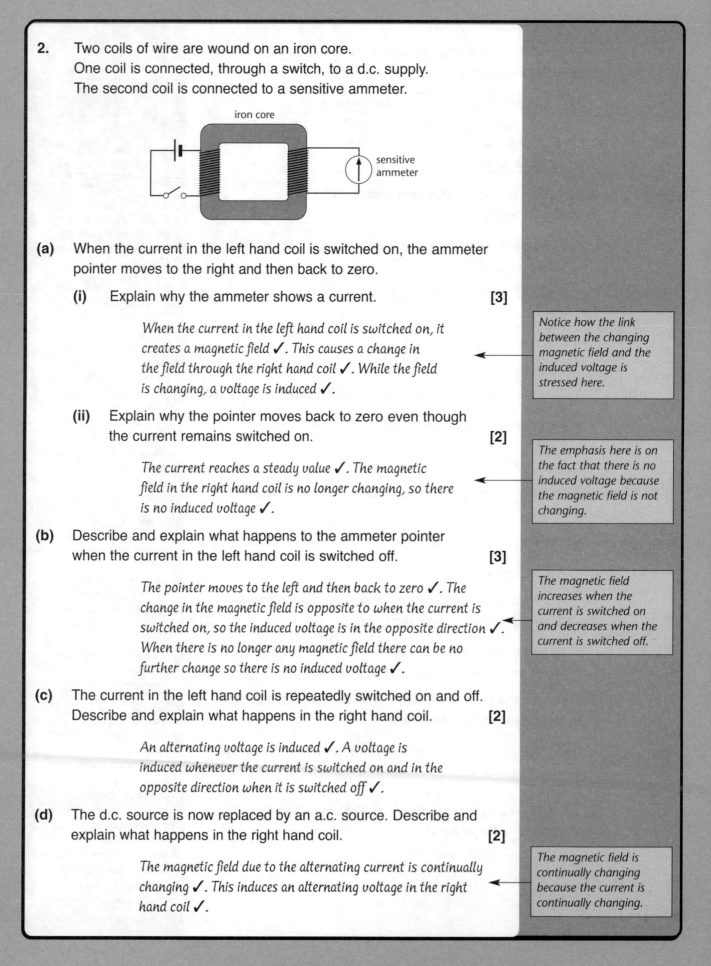

iron core

sensitive ammeter

(a) When the current in the left hand coil is switched on, the ammeter pointer moves to the right and then back to zero.

    (i) Explain why the ammeter shows a current. [3]

       When the current in the left hand coil is switched on, it creates a magnetic field ✓. This causes a change in the field through the right hand coil ✓. While the field is changing, a voltage is induced ✓.

> Notice how the link between the changing magnetic field and the induced voltage is stressed here.

    (ii) Explain why the pointer moves back to zero even though the current remains switched on. [2]

       The current reaches a steady value ✓. The magnetic field in the right hand coil is no longer changing, so there is no induced voltage ✓.

> The emphasis here is on the fact that there is no induced voltage because the magnetic field is not changing.

(b) Describe and explain what happens to the ammeter pointer when the current in the left hand coil is switched off. [3]

       The pointer moves to the left and then back to zero ✓. The change in the magnetic field is opposite to when the current is switched on, so the induced voltage is in the opposite direction ✓. When there is no longer any magnetic field there can be no further change so there is no induced voltage ✓.

> The magnetic field increases when the current is switched on and decreases when the current is switched off.

(c) The current in the left hand coil is repeatedly switched on and off. Describe and explain what happens in the right hand coil. [2]

       An alternating voltage is induced ✓. A voltage is induced whenever the current is switched on and in the opposite direction when it is switched off ✓.

(d) The d.c. source is now replaced by an a.c. source. Describe and explain what happens in the right hand coil. [2]

       The magnetic field due to the alternating current is continually changing ✓. This induces an alternating voltage in the right hand coil ✓.

> The magnetic field is continually changing because the current is continually changing.

Sample IGCSE questions

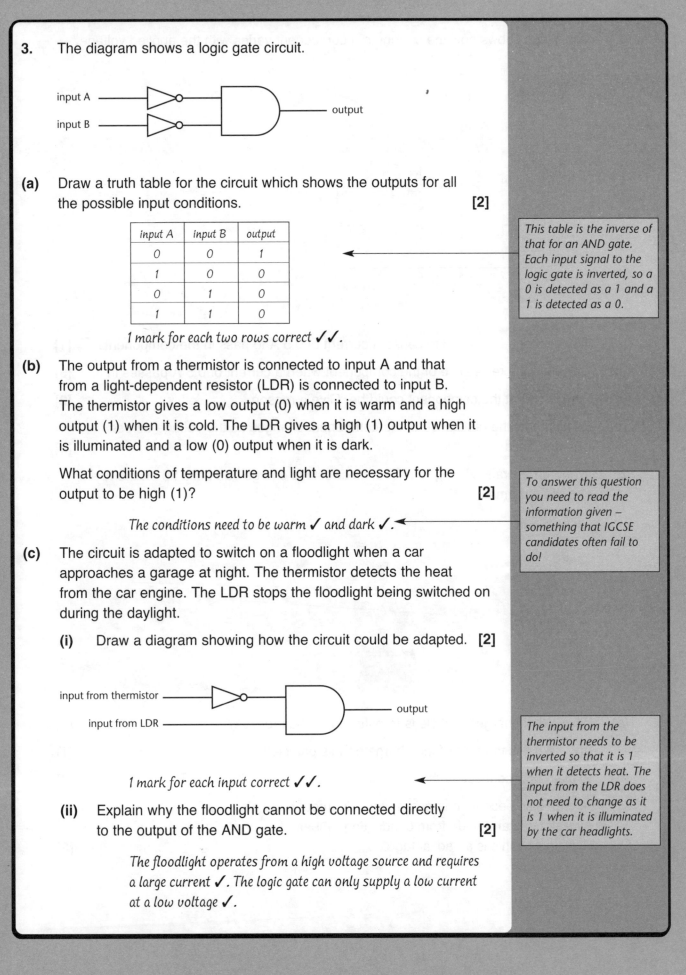

3. The diagram shows a logic gate circuit.

input A

input B

output

(a) Draw a truth table for the circuit which shows the outputs for all the possible input conditions. [2]

input A	input B	output
0	0	1
1	0	0
0	1	0
1	1	0

1 mark for each two rows correct ✓✓.

> This table is the inverse of that for an AND gate. Each input signal to the logic gate is inverted, so a 0 is detected as a 1 and a 1 is detected as a 0.

(b) The output from a thermistor is connected to input A and that from a light-dependent resistor (LDR) is connected to input B. The thermistor gives a low output (0) when it is warm and a high output (1) when it is cold. The LDR gives a high (1) output when it is illuminated and a low (0) output when it is dark.

What conditions of temperature and light are necessary for the output to be high (1)? [2]

The conditions need to be warm ✓ and dark ✓.

> To answer this question you need to read the information given – something that IGCSE candidates often fail to do!

(c) The circuit is adapted to switch on a floodlight when a car approaches a garage at night. The thermistor detects the heat from the car engine. The LDR stops the floodlight being switched on during the daylight.

(i) Draw a diagram showing how the circuit could be adapted. [2]

input from thermistor

input from LDR

output

1 mark for each input correct ✓✓.

> The input from the thermistor needs to be inverted so that it is 1 when it detects heat. The input from the LDR does not need to change as it is 1 when it is illuminated by the car headlights.

(ii) Explain why the floodlight cannot be connected directly to the output of the AND gate. [2]

The floodlight operates from a high voltage source and requires a large current ✓. The logic gate can only supply a low current at a low voltage ✓.

Exam practice questions

1. The graph shows how the current in a component varies with the applied voltage.

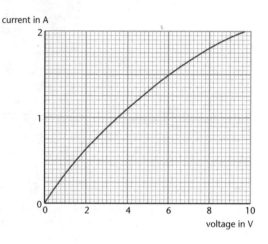

(a) What voltage is needed to cause a current of 0.5 A to pass in the component? **[1]**

(b) Calculate the resistance of the component when a current of 0.5 A passes in it. **[3]**

(c) Suggest what the component could be. Give a reason. **[2]**

(d) Explain how the resistance of the component changes as the voltage across it is increased. **[2]**

2. When an aircraft is being refuelled, electrostatic charge can build up on the fuel and the airframe.

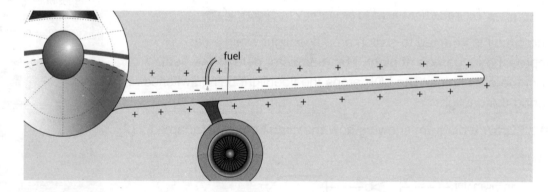

(a) (i) What charged particle is transferred in this process? **[1]**

(ii) What is the sign of the charge on this particle? **[1]**

(b) Explain how connecting the airframe to earth prevents the build up of charge. **[3]**

(c) Aircraft also become charged as they fly through the air.
Aircraft tyres are made from conducting rubber.
Explain why this is an advantage. **[3]**

Exam practice questions

3. This question is about a car headlamp operating from a 12 V supply.

(a) 300 C of charge pass through the filament in one minute.

Calculate the current in the lamp. [3]

(b) How much energy is transferred to the filament by this charge passing through it? [3]

(c) Calculate the power of the lamp. [3]

4. An electric shower has a power of 8.4 kW when it is connected to the 240 V mains supply.

(a) Calculate the current in the heater when it is operating normally. [3]

(b) Explain why very thick cables are needed to connect it to the mains supply. [2]

(c) It is recommended that a residual current circuit breaker (RCCB) is placed in the live supply connection.

What are the advantages of using an RCCB instead of a fuse? [2]

(d) The shower is used for 30 minutes each day for the week.

Each kW h of energy costs 7p.

Use the relationship:

energy in kW h = power in kW × time in h

to calculate the cost of using the shower each week. [3]

5. Here are three graphs that show how the current in a component changes when the voltage is changed.

(a) Which graph represents a filament lamp? [1]

(b) Which graph represents a fixed resistor at a constant temperature? [1]

(c) Which graph represents a component that only allows current to pass in one direction? [1]

(d) Which graph represents a diode? [1]

(e) The current in a lamp filament is 2.5 A when the voltage across it is 12.0 V.

Calculate the power of the lamp. [3]

Exam practice questions

6. A voltage is induced in a conductor when the magnetic field through it changes.

(a) What is the name of this effect? [1]

(b) The diagram shows a coil of wire next to a magnet. A voltmeter is connected to the coil of wire.

(i) Describe TWO ways of inducing a voltage in the coil of wire. [2]

(ii) State THREE factors that affect the size of the induced voltage. [3]

(iii) Write down TWO ways of changing the direction of the induced voltage. [2]

(c) A transformer consists of two coils of wire on an iron core. An alternating voltage applied to the input (primary) coil causes an alternating voltage in the output (secondary) coil.

(i) A 3 V battery is connected to the input coil.
Explain why there is no voltage across the output coil. [2]

(ii) A 3 V a.c. source is connected to the input coil.
Explain why there is a voltage across the output coil. [2]

(iii) The input coil has 200 turns.
When the 3 V a.c. source is connected to this coil, the output voltage is
12 V. Calculate the number of turns on the output coil. [2]

7. A power station generator consists of an electromagnet that rotates inside three sets of copper conductors. The current in each conductor is 7500 A.

(a) Water flows through channels inside the conductors.
Suggest why this is necessary. [2]

(b) Electricity is transmitted along the national grid using a combination of overhead and underground cables.

(i) Suggest TWO reasons why overhead conductors are used in preference to underground conductors outside towns and cities. [2]

(ii) Suggest TWO reasons why underground conductors are used in preference to overhead conductors in towns and cities. [2]

(c) Explain why renewable sources of energy are used to produce only a small proportion of the electricity generated in the United Kingdom. [3]

Exam practice questions

8. (a) Explain how the construction of a step-up transformer differs from that of a step-down transformer. [3]

(b) Describe how transformers are used in the distribution of electricity. [2]

(c) Explain why it is necessary to distribute electricity at high voltage. [2]

(d) Explain why mains electricity is alternating current rather than direct current. [2]

9. Electricity can be generated from renewable and non-renewable energy resources.

(a) Name TWO renewable energy resources. [2]

(b) Name TWO non-renewable energy resources. [2]

(c) Explain how gas-fired power stations cause less damage to the environment than coal-fired power stations. [3]

(d) Suggest why it is a social advantage to burn British coal in coal-fired power stations. [2]

(e) (i) Give ONE advantage and TWO disadvantages of generating electricity from wind power. [3]

(ii) Suggest why very little electricity is generated in the UK from moving water. [3]

(iii) France has limited fossil fuel reserves, mostly coal. Most of its electricity is generated from nuclear power. Outline the advantages and disadvantages of generating electricity from nuclear power. [4]

10. The diagram shows a d.c. motor.

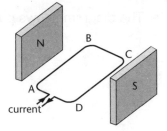

(a) Draw an arrow on the diagram that shows the direction of the magnetic field. [1]

(b) (i) Explain why, in the position shown, the sides AB and CD of the coil experience a force. [1]

(ii) Explain why there is no force on the sides BC and AD. [1]

(c) (i) Explain why the forces on the coil cause it to rotate. [2]

(ii) State TWO ways in which the direction of rotation can be reversed. [2]

(iii) State TWO ways in which the speed of rotation can be increased. [2]

Exam practice questions

11. The diagram shows a logic gate circuit that is part of a car.

It is designed to operate an alarm if either the driver's or passenger's door is opened while the headlights are switched on.

(a) Write down the names of the logic gates used in this circuit. [2]

(b) Complete the truth table for the circuit. [4]

input A	input B	input C	output
0	0	0	
1	0	0	
0	1	0	
1	1	0	
0	0	1	
1	0	1	
0	1	1	
1	1	1	

(c) Which input should be connected to the headlights? [1]

(d) Explain whether the alarm is on when both doors are opened with the headlights switched on. [2]

12. The heater in an incubator is controlled by a logic gate. The diagram shows part of the circuit used.

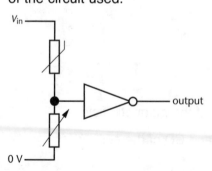

(a) Explain how the heater is switched on when the temperature of the incubator falls. [3]

(b) What is the advantage of using a variable resistor rather than a fixed resistor? [1]

Exam practice questions

(c) The input voltage to the circuit is 5.0 V.

At a temperature of 35°C the resistance of the thermistor is 45.0 Ω.

The logic gate switches when the input voltage falls to 2.0 V.

Calculate the resistance of the variable resistor if the heater switches on when the temperature drops to 35°C. [3]

13. A commercial freezer is fitted with an alarm.

The alarm sounds if the door is left open or the temperature exceeds −10°C.

The diagram shows the system.

(a) Which part of this system is:

(i) the processor? [1]

(ii) the output device? [1]

(b) The logic gate used is an OR gate.

Complete the truth tables for the door switch and the temperature sensor. [2]

door switch	
input conditions	output
door open	
door closed	

temperature sensor	
input conditions	output
less than −10°C	
more than −10°C	

(c) The OR gate is replaced with a NOR gate.

(i) Under what conditions does the alarm now sound? [2]

(ii) Explain how the circuit can be adapted so that it functions as before. [2]

14. The diagram shows a potential divider circuit.

Exam practice questions

(a) Explain how the output voltage changes when the illumination of the LDR decreases. **[2]**

(b) The input voltage to the circuit is 5.0 V.

The output from the LDR is 3.0 V when its resistance is 2000 Ω.

Calculate the resistance of the fixed resistor. **[3]**

(c) Explain how the output voltage is affected by increasing the resistance of the fixed resistor. **[2]**

15. The diagram shows a circuit that consists of four resistors and a power supply.

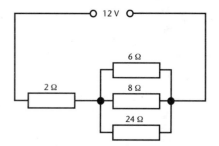

(a) Calculate the effective resistance of the three resistors in parallel. **[1]**

(b) Complete the table. **[4]**

resistor	current (A)	voltage (V)
2 Ω		
6 Ω		
8 Ω		
24 Ω		

16. The diagram shows a device for producing and accelerating electrons.

Exam practice questions

(a) Name the process that results in the release of electrons from the filament. [1]

(b) Explain why the filament needs to be hot to emit electrons. [2]

(c) Explain how the electrons are accelerated. [2]

(d) The anode voltage is 2500 V.
The electronic charge, $e = 1.6 \times 10^{-19}$ C.
Calculate the kinetic energy of the electrons in the beam. [2]

17. The diagram shows an oscilloscope trace.

y – sensitivity = 2 V/cm
time base = 1 ms/cm

(a) How can you tell from the diagram that the input to the oscilloscope is an alternating voltage [1]

(b) (i) What is the maximum value of the input voltage? [1]

(ii) What is the minimum value of the input voltage? [1]

(c) Calculate the frequency of the input voltage. [3]

5 Atomic and particle physics

5.1 Ionising radiations

LEARNING SUMMARY

After studying this section you should be able to:

● distinguish between the three main radioactive emissions in terms of their penetration and ionisation
● describe the effects of radiation on the body
● explain the effect on the nucleus when it decays by emitting radiation.

Radiation from the nucleus

If a material is **radioactive** its atomic nuclei are **unstable**; when they change to a more stable form they emit **radiation**. Some materials are naturally radioactive, others become radioactive when the nucleus absorbs neutrons, as happens in a nuclear reactor. The three main types of nuclear radiation are:

● **alpha (α)** – this is intensely ionising but has a very low penetration, being absorbed by a few centimetres of air or a sheet of paper

● **beta (β)** – this is strongly ionising and more penetrative than alpha radiation. It is partly absorbed by thick paper or card and completely absorbed by a few millimetre thickness of aluminium or other metal

> Beta radiation causes more ionisation in the materials it passes through than gamma radiation, but less than alpha radiation.

● **gamma (γ)** – this is only weakly ionising but very penetrative; its intensity is reduced by using thick lead or concrete walls.

KEY POINT

● An alpha particle consists of two protons and two neutrons
● A beta particle is a fast-moving electron
● Gamma radiation is a high-frequency, short-wavelength electromagnetic wave.

All three types of radiation can be detected by a **Geiger-Müller** tube and by **photographic film**, which blackens by exposure to radiation in the same way as it does when it is exposed to light.

Radiation all around us

We are constantly being bombarded by radiation from our surroundings, called the **background radiation**. A radioactive form of carbon, **carbon-14**, is created in the atmosphere and absorbed by plants, so radioactivity is present throughout all food chains. Background radiation is due to:

> Everything we eat depends on plants, so radioactive carbon absorbed by plants is present in all our food.

- radioactivity in all plants and animals

- radiation from the Sun and space

- radiation from buildings

- radiation from the ground

- radiation from hospitals and industrial users of radioactive materials

- radiation from waste materials and "leaks" from nuclear power stations.

highest level

medium

lowest level

Fig. 5.1

> **Granite contains radium, which decays by the emission of an alpha particle to form radon, a radioactive gas which seeps out of the rock.**

Granite is a radioactive rock that emits a radioactive gas called **radon**. This gas accumulates in buildings and is a particular hazard when breathed in, as it emits **alpha** radiation. The occurrence of granite is one reason why the level of background radiation is higher in some parts of the country than in others. The variation in the levels of background radiation is shown in the diagram, **Fig. 5.1**.

Radiation can affect the body in a number of ways:

- it can destroy **cells** and **tissue**

- it can also change the DNA, causing **mutations** and affecting future generations if the sex cells are affected

- radiation can cause **skin burns** and **cancer**.

Alpha radiation is particularly hazardous when it enters the body because of its ability to damage cells. It is less of a threat when outside the body because of its low penetration.

Beta radiation is a hazard both inside the body and from the outside. Its penetration allows it to pass through the skin and be absorbed by body tissue, where its **ionising ability** enables it to cause damage to cells and tissue.

> **In radiotherapy, gamma emitters target specific areas of the body. Cancer cells are affected more than normal cells because they reproduce more rapidly.**

Although it is the most penetrative radioactive emission, **gamma** radiation is less hazardous than beta radiation because it is less likely to be absorbed by body tissue. When it is absorbed, it can destroy cells and is used in radiotherapy treatment to kill cancer cells.

To reduce the hazards from radioactive materials:

- people should be **shielded** from radioactive sources by a suitable absorber

- there should be **as large a distance** as possible between a person and a radioactive source

- the time of exposure to radiation should be as **short** as possible

- people who work with radioactive materials wear a badge containing photographic film that measures the amount of exposure to radiation.

The effect on the nucleus

The scattering of alpha particles by thin gold foil shows that the atom is mainly empty space, with a **large concentration** of **mass** and **positive charge** in a tiny volume. This is the atomic **nucleus**, which is surrounded by orbiting electrons. The nucleus contains two types of particle:

> Alpha particle scattering experiments were first carried out by Geiger and Marsden under the guidance of Lord Rutherford.

- **neutrons,** which have no charge

- **protons,** which have the same mass as neutrons and carry a single **positive** charge.

The orbiting **electrons** have very little mass and each carries a single **negative** charge. A neutral atom contains equal numbers of protons and electrons.

> **KEY POINT**
> the atomic number or proton number (**Z**) is the number of protons in the nucleus
> the mass number or nucleon number (**A**) is the total number of (protons + neutrons) in the nucleus.

The structure of a nucleus is represented in symbol form as $^{A}_{Z}\text{El}$, for example $^{24}_{12}\text{Mg}$ represents an atom of magnesium that has twelve protons and twelve neutrons.

The number of protons in the nucleus determines the element, but not all atoms of the same element are identical. Some elements exist in different forms called **isotopes**.

> Hydrogen has two isotopes, called deuterium and tritium. Unlike hydrogen, the nuclei contain neutrons; one in the case of deuterium and two in the case of tritium.

> **KEY POINT**
> Isotopes of an element have the same number of protons but different numbers of neutrons in the nucleus.

The isotopes of an element all have the same chemical properties but vary in the physical properties of density and nuclear stability.

When a radioactive isotope decays by alpha or beta emission, it results in the formation of the nucleus of a different element:

- **alpha** emission causes the loss of **two protons** and **two neutrons**, so the nucleus formed has an atomic number which has decreased by two and a mass number which has decreased by four

- in **beta** emission a **neutron** decays to a **proton** and an **electron**; the atomic number increases by one and the mass number is unaffected

> In beta decay the electron is ejected from the nucleus, leaving it with one more proton and one less neutron.

- **gamma** emission does not affect the mass or atomic number; it is emitted to reduce the excess energy of the nucleus, often following alpha or beta emission.

The equations that represent radioactive decay are balanced in terms of charge (represented by Z, the number of protons) and mass (represented by A, the number of nucleons). The decay of carbon-14 by beta-emission and radon-220 by alpha-emission are represented by the equations:

> Check the balance of these equations. The upper numbers represent mass and the lower numbers represent charge.

$$^{14}_{6}\text{C} \rightarrow {}^{14}_{7}\text{N} + {}^{0}_{-1}\text{e}$$

$$^{220}_{86}\text{Rn} \rightarrow {}^{216}_{84}\text{Po} + {}^{4}_{2}\alpha$$

PROGRESS
CHECK

1. Which type of radioactive emission:
   (a) Is the most intensely ionising?
   (b) Has the greatest penetration?
2. Alpha-emitters are used in smoke alarms. Explain why these present no danger to people.
3. Which type of radioactive emission has no effect on the mass number or atomic number of the atom?

1(a) Alpha; (b) Gamma; 2. The alpha particles cannot penetrate the plastic casing; 3. Gamma.

5.2 Using radiation

LEARNING
SUMMARY

After studying this section you should be able to:

● *explain how the decay of a radioactive isotope changes with time*
● *recall and use the term half-life*
● *describe how radioactive isotopes are used to date rocks and other objects.*

Radioactive decay and half-life

Radioactive decay is a **random** process; the decay of any particular nucleus is unpredictable and, unlike chemical reactions, radioactive decay is not affected by physical conditions such as temperature. The rate at which an **isotope** decays depends on:

● the number of **undecayed nuclei** present in the sample; on average, doubling the number of undecayed nuclei should double the rate of decay

● the **stability** of the radioactive isotope; some isotopes decay much more rapidly than others.

KEY
POINT

> The rate of decay is the number of nuclei that decay each second.
> It is measured in becquerel (Bq) where 1 Bq = 1 decay/s

The rate of decay is proportional to the number of undecayed nuclei present in the sample.

As a sample of a radioactive isotope decays, the number of undecayed nuclei decreases, and so the **rate of decay** also decreases. The graph, **Fig. 5.2** shows a typical decay curve.

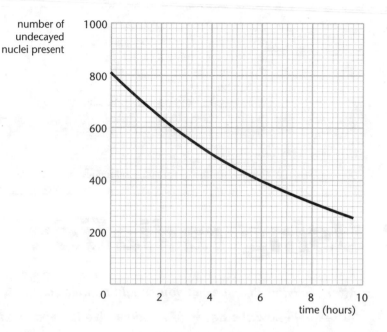

A graph of rate of decay against time would have the same shape as this, the only difference being the figures on the *y*-axis.

number of undecayed nuclei present

Fig. 5.2

This graph shows that the time it takes for the number of undecayed nuclei to **halve** from 800 to 400 is the same as the time it takes for the number of undecayed nuclei to **halve** from 600 to 300 or between any other two corresponding values. The actual time this takes varies from isotope to isotope but for any one isotope it has a fixed value.

> **KEY POINT**
> The half-life of a radioactive isotope is the average time it takes for the number of undecayed nuclei to halve.

Because it is difficult to measure the number of undecayed nuclei in a sample, half-life is often measured as the time it takes for the rate of decay to halve.

So, after one half-life, half of the original undecayed nuclei would be expected to remain with one quarter being left after two half-lives have passed.

> **KEY POINT**
> After *n* half-lives have elapsed, $\frac{1}{2_n}$ of the original undecayed nuclei present in a sample remain. The rest have changed into nuclei of a different element.

Radioactive dating

For a more reliable measurement, a mass spectrometer is used to compare the amounts of carbon-12 and carbon-14.

All **living things** are continually absorbing **radioactive carbon-14** from the air or their food. The concentration of carbon-14 in a plant or animal stays at a constant level until it dies. After that it decreases as the carbon-14 decays with a **half-life** of 5730 years and no fresh carbon-14 is absorbed. Archaeological specimens of once-living material such as wood can be dated by measuring the activity of the radioactive carbon remaining.

When **igneous rocks** are formed from **magma**, they contain uranium-238, which decays to form lead-206. The half-life of this decay is 4500 million years. The age of a rock can be dated by comparing the amounts of lead-206 and uranium-238 that it contains. For the oldest rocks found, these isotopes exist in approximately equal amounts, putting the age of the Earth at about 4500 million years.

A similar technique is used for rocks formed containing potassium-40. This has a half-life of 1300 million years; it decays to form argon which can become trapped in the rock. This method can be unreliable because there is no certainty that some of the argon has not seeped out of the rock.

Nuclear power

Energy is released when fission of large nuclei takes place. This is a process in which they are broken up into a number of smaller particles. The fission of uranium-235 is shown in the diagram, **Fig. 5.3.**

> This is a different reaction to that which takes place in stars where energy is released when small nuclei fuse together.

BEFORE
uranium
neutron

AFTER
barium
neutron
neutron
krypton
neutron

Fig. 5.3

In this process:

- a uranium-235 nucleus absorbs a **neutron**, making it **unstable**
- it splits into two smaller nuclei, which are also unstable, and two or three spare neutrons
- the fission products have a lot of **kinetic energy**, which is removed by a coolant
- the coolant generates steam that turns a turbine
- the **spare neutrons** can cause **further fission** of other nuclei.

> In the core of a nuclear reactor, control rods absorb the spare neutrons to control the rate of the reaction.

If each spare neutron were allowed to cause another fission, the result would be a chain reaction which would be out of control. To maintain the reaction at a steady rate, on average just one of the neutrons released by each fission is allowed to go on and cause a further fission.

Nuclear power has one major disadvantage – how to get rid of the waste materials. These are in three categories:

- low-level waste such as laboratory clothing and packaging materials; these are buried either underground or at sea
- intermediate-level waste such as the casing used for nuclear fuel and reactor parts that have been replaced; these are kept in stores with thick concrete walls or buried in deep trenches with concrete linings
- high-level waste such as spent fuel rods; these present a long-term disposal problem since they remain significantly radioactive for thousands of years; much of this waste is in temporary storage in tanks of water until the problem of what to do with it can be solved.

Some other uses of radioactivity

Radioactive isotopes are also used as **tracers** in medicine and to control the thickness of sheet materials.

When used as a **tracer**:

● the **half-life** of the isotope used should be **long** enough for the tracer still to be radioactive when it has reached its target

● the **half-life** should be **short** enough so that the patient does not remain radioactive for a long period of time, causing unnecessary risk

● the isotope should emit **gamma** radiation **only**; this can be detected outside the body whereas alpha and beta radiations would be strongly absorbed, causing cell damage.

The diagram, **Fig. 5.4**, shows how a radioactive isotope is used to control the thickness of sheet materials.

Fig. 5.4

For many sheet materials a beta source is the most suitable.

In this application:

● the radiation from the isotope used must be **partially absorbed** by the material

● if the thickness of the sheet increases, less radioactivity is detected; this information is fed back and the pressure on the rollers is increased

● the isotope should have a **long half-life** to avoid the need for constant recalibration.

PROGRESS CHECK

1. The rate of decay of a sample of a radioactive isotope depends on two factors. What are these factors?
2. A radioactive isotope with a half-life of 6 hours is used as a tracer in medicine. What fraction of the original nuclei remain after one day?
3. Why is a beta-emitter the most suitable for controlling the thickness of paper?

1. The number of undecayed nuclei and the particular isotope in the sample;
2. 1/16;   3. Alpha radiation would be totally absorbed by the paper. No gamma radiation would be absorbed by the paper.

5.3 Atoms and nuclei

LEARNING SUMMARY

After studying this section you should be able to:

● *explain how the results of alpha particle scattering give evidence for the atomic model*
● *explain whether an isotope is stable or unstable, by reference to the stability curve*
● *describe the changes in the nucleus when an isotope undergoes beta decay.*

Farewell to the plum pudding

Prior to the results of experiments carried out by Geiger and Marsden, who worked under the guidance of Lord Rutherford in 1911, the atom was thought to have the structure of a "**plum pudding**". The negatively-charged electrons were pictured as being evenly distributed within a positively-charged uniform mass.

> The alpha particles were detected by the flashes of light emitted when they hit a fluorescent screen.

Geiger and Marsden fired **alpha particles** from radon gas at thin gold foil. They then detected the alpha particles after they had been scattered. They found that:

● most of the alpha particles pass through the foil with no deflection
● some are deflected through a range of angles
● a small number are "back-scattered".

These outcomes are shown in the diagram, **Fig. 5.5**.

> The proportion of the alpha particles that are "back-scattered" is about 1 in 10 000.

gold atom

alpha particles

gold foil

Fig. 5.5

Rutherford concluded from these results that:

● the atom is mainly empty space, allowing most of the alpha particles to pass through undeflected
● there are tiny regions of concentrated charge, which explains the large deflection of a small number of the alpha particles
● the charge on these regions must be the same sign as that on an alpha particle – positive – to cause the "**back-scattering**".

The currently accepted atomic model, shown in the diagram, **Fig. 5.6**, is that the atom has a central **nucleus** which contains almost all of the mass of the atom but occupies a very small fraction of the atomic volume. The nucleus has a positive charge which is balanced by the negative charge of orbiting electrons, so that overall the atom is neutral.

volume around the nucleus containing electrons **Fig. 5.6**

protons and neutrons packed together in the nucleus (positively charged)

> In scattering experiments the electrons do not affect the alpha particles, because of the vast difference in their masses.

Alpha particle-scattering experiments carried out using other materials show that the amount by which an alpha particle is deflected depends on:

- the closeness of its approach to the nucleus – the closer it gets, the bigger the force
- the charge on the nucleus – the greater the charge, the greater the force of repulsion
- the speed of the alpha particle – the faster it travels, the smaller the deflection.

Nuclear stability

Whether or not a nucleus is **stable** depends on the balance of protons and neutrons. The graph, **Fig. 5.7**, shows the relationship between the number of neutrons (N) and the number of protons (Z) for stable nuclei. Isotopes with nuclei in the regions of instability can decay in a number of ways according to their position on the graph:

regions of instability

neutrons = protons

stability line

number of protons (Z)

Fig. 5.7

- isotopes that are above the line have too many neutrons to be stable – they decay by β^- **emission** when a neutron emits an electron, changing to a proton

> A positron is identical to an electron, but it has the opposite charge. A positron is very short-lived as it is annihilated when it meets an electron, resulting in two or three gamma ray photons.

- isotopes that are below the line have too many protons to be stable – they decay by β^+ **emission** when a proton emits a positron (an anti-electron), changing to a neutron
- isotopes with more than 82 protons in the nucleus usually decay by **alpha-emission**.

When a nucleus decays by emitting an alpha or beta particle the new nucleus that is formed often has too much energy to be stable. It loses this excess energy by emitting gamma radiation.

How does beta decay affect the nucleus?

Electrons are **fundamental particles**, they are not made up of any other, smaller particles. This is not the case for protons and neutrons – each is made up of three smaller particles called **quarks**. There are a number of different types of quark, but only two are used to build protons and neutrons, the **up** quark and the **down** quark. The diagram, **Fig. 5.8**, shows the structure of a neutron and a proton and the charges on these quarks.

> Quarks cannot exist in isolation and they themselves may be made up of other particles.

up $q=+\frac{2}{3}e$ down $q=-\frac{1}{3}e$

neutron   down $q=-\frac{1}{3}e$

up $q=+\frac{2}{3}e$ up $q=+\frac{2}{3}e$

proton   down $q=-\frac{1}{3}e$

Fig. 5.8

In β^- decay:

● a down quark in a neutron changes to an up quark, emitting an electron

● the atomic number (number of protons, Z) increases by one

● the mass number (total number of nucleons, A) is unchanged.

An example of β^- decay is the decay of strontium-90 into yttrium-90:

$$^{90}_{38}\text{Sr} \rightarrow ^{90}_{39}\text{Y} + ^{0}_{-1}\text{e}$$

Check that these equations are balanced in terms of mass and charge.

In β^+ decay:

● an up quark in a proton changes to a down quark, emitting a positron

● the atomic number decreases by one

● the mass number is unchanged.

An example of β^+ decay is the decay of carbon-11 to boron-11:

$$^{11}_{6}\text{C} \rightarrow ^{11}_{5}\text{B} + ^{0}_{+1}\text{e}$$

PROGRESS CHECK

1. How does "back-scattering" show that the atomic nucleus has the same sign as that on an alpha particle?
2. Explain why isotopes that lie above the stability line in fig 5.7 are sometimes described as being "neutron-rich".
3. Why is an electron called a fundamental particle?

1. In back-scattering the alpha particle is repelled by the nucleus. Objects with similar charges repel each other;   2. They have too many neutrons to be stable;   3. The electron is not made up of any other particles.

Sample IGCSE questions

1. $^{12}_{6}$C and $^{14}_{6}$C are both isotopes of carbon.

(a) **(i)** Write down one similarity about the nucleus of each isotope. **[1]**

> *They have the same number of protons* ✓.

(ii) Write down one difference in the nucleus of these isotopes. **[1]**

> *They have different numbers of neutrons* ✓.

(b) $^{14}_{6}$C is radioactive. It decays by emitting a beta particle.

(i) Describe a beta particle. **[1]**

> *A beta particle is a fast-moving electron* ✓.

(ii) Which part of the atom emits the beta particle? **[1]**

> *The nucleus* ✓. ←————————————

A common error at IGCSE is to state that the beta particle comes from the electrons that orbit the nucleus, since the nucleus does not contain any electrons.
All nuclear radiation is emitted from the nucleus; in this case a neutron decays to an electron and a proton.

(c) $^{14}_{6}$C is present in all living materials and in all materials that have been alive. It decays with a half-life of 5730 years. **[2]**

(i) Explain the meaning of the term *half-life*.

> *Half-life is the average time* ✓ *for the number of undecayed nuclei to halve* ✓.

(ii) The activity of a sample of wood from a freshly-cut tree is measured to be 80 Bq. Estimate the activity of the sample after two half-lives have elapsed. **[1]**

> *20 Bq* ✓. ←————————————

A common misunderstanding is that after two half-lives all the nuclei have decayed. This is not the case; on average one half of one half, ie one quarter, of the original nuclei are undecayed after two half-lives.

(iii) The age of old wood can be estimated by measuring its radioactivity. Explain why this method cannot be used to work out the age of a piece of furniture made in the nineteenth-century. **[2]**

> *One to two hundred years is a very short time compared to the half-life* ✓. *The rate of decay would not show any significant change* ✓.

(iv) Explain why radiocarbon dating cannot be used to estimate the age of a rock. **[2]**

> *Radiocarbon dating works by measuring the decay of carbon-14, which is found in all living things* ✓. *Rock has never lived, so it does not contain any carbon-14* ✓.

Sample IGCSE questions

2. (a) The diagram shows the "plum pudding" model of the atom.

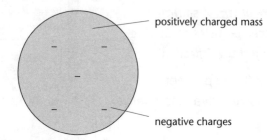

positively charged mass

negative charges

Explain how the position of the electrons on the currently accepted atomic model is different to that on the "plum pudding" model. **[2]**

It is important here to use information given in the diagram. The diagram shows the electronic arrangement in the "plum pudding" model. The first mark is for describing the arrangement shown.

> *On the "plum pudding" model the electrons are positioned throughout the atom ✓ but on the currently accepted model they orbit the nucleus ✓.*

(b) The next diagram shows some of the results when alpha particles are fired at thin gold foil.

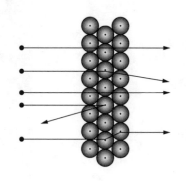

(i) Explain how this shows that the mass of the atom is not distributed uniformly. **[2]**

The cue word "explain" means that you need to give full reasons to support your answers. In this case the reasons are the different paths of the alpha particles that pass through the foil.

> *Many alpha particles pass straight through, so they cannot have encountered any of the atomic mass ✓. Other particles are deviated by different amounts, so they cannot have passed through a uniformly distributed mass ✓.*

(ii) Explain how this gives evidence about the charge on the atomic nucleus. **[2]**

> *The charge on the nucleus must be the same as that on an alpha particle ✓ since some alpha particles are repelled ✓.*

Sample IGCSE questions

(iii) Suggest reasons why some of the particles that pass through the foil are deflected more than others. **[3]**

Some pass closer to the nucleus than others ✓ so they experience a greater force ✓. The alpha particles could have different speeds, which would also affect the deflection ✓.

(iv) Suggest how the results might differ if the experiment is carried out using a metal foil whose nuclei have less charge than those of gold. **[2]**

There would be less deflection overall ✓ since the repulsive force would be smaller at the same distance ✓.

> The cue word "suggest" means that this may not be required knowledge, but the question requires you to apply your knowledge and understanding of other areas of physics to this particular situation.

Exam practice questions

1. (a) The three main types of radioactive emission are called alpha, beta and gamma.

Alpha particles are positively charged.
They consist of two protons and two neutrons.
They are absorbed by thin paper.

Write similar descriptions of beta particles and gamma radiation. **[6]**

(b) A source of gamma radiation is pointed at a Geiger-Müller tube connected to a ratemeter.

Sheets of lead of different thicknesses are placed between the source and the Geiger-Müller tube.

The count rate is measured for each lead sheet. The results are shown in the table.

thickness of lead sheet (mm)	0	5	10	15	20	25
count rate (Bq)	76	57	43	32	24	18

(i) Use a grid to draw a graph of count rate against thickness. **[4]**

(ii) What thickness of lead is needed to halve the intensity of the gamma radiation? Explain how you obtain your answer. **[2]**

(iii) A similar source emits gamma rays at the rate of 120/s. It is to be transported in a lead container. The emission of gamma rays from the container must not exceed 15/s.

What minimum thickness of container is required? **[3]**

2. Technetium-99 is a radioactive isotope used as a tracer in medicine. It decays by emitting gamma radiation only with a half-life of six hours. It is injected into the bloodstream and detected using a camera placed outside the body.

(a) Explain the meaning of the terms isotope and half-life. **[4]**

(b) (i) Why is it important that the isotope used for this purpose emits gamma radiation? **[2]**

(ii) Why is it desirable that the isotope used does not emit alpha or beta radiation? **[2]**

(c) Explain why 6 hours is a suitable half-life for an isotope used as a medical tracer.

3. A living tree is radioactive. After it dies, the decay rate of carbon-14 in the tree decreases.

(a) (i) Suggest why the decay rate does not decrease when the tree is alive. **[1]**

(ii) Explain why the decay rate decreases when the tree has died. **[2]**

Exam practice questions

(b) The graph shows how the decay of the carbon-14 in a 0.5 kg sample of wood from the tree changes after the tree has died.

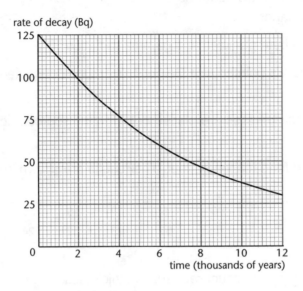

rate of decay (Bq)

time (thousands of years)

(i) Estimate the decay rate of a 2 kg sample of wood from a tree immediately after it has died. **[1]**

(ii) Use the graph to calculate the half-life of carbon-14.
Show how you obtain your answer. **[2]**

(iii) A 0.10 kg sample of wood from an archaeological excavation is found to have a decay rate of 12.0 Bq.

Use the graph to estimate the age of the wood. **[2]**

4. Sodium-24 is a radioactive form of sodium that emits gamma radiation. It has a half-life of 15 hours.

In the form of sodium chloride, it is used to detect leaks from underground water pipes.

(a) Suggest two reasons why sodium-24 is a suitable isotope to use for this purpose. **[2]**

(b) Suggest how it should be used. **[2]**

(c) **(i)** What could be used to detect radiation from the water? **[1]**

(ii) How would a person operating the detector be able to tell where water was leaking from the pipe? Give the reason for this. **[2]**

(d) The water is safe to drink when the radioactivity is one eighth of its initial value. What minimum time should elapse before anyone drinks the water? **[2]**

Exam practice questions

5. The graph shows the relationship between the number of protons and the number of neutrons in stable nuclei.

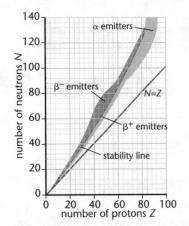

(a) Explain why no gamma emitters are shown on the graph. **[2]**

(b) Explain what type of radiation is emitted by the isotope strontium-90, $^{90}_{38}$Sr. **[3]**

(c) Complete the equation for the decay of strontium-90. **[2]**

$$^{90}_{38}\text{Sr} \rightarrow \text{Y}$$

(d) The stability of a nucleus depends on the neutron-proton ratio.

   (i) Explain how the neutron-proton ratio for stable nuclei changes as the proton number increases. **[3]**

   (ii) Suggest why there are no alpha emitters with less than ten protons in the nucleus. **[2]**

The Earth and beyond

The following topics are covered in this section:

● *The Solar System and its place in the Universe*   ● *Evolution*

6.1 The Solar System and its place in the Universe

LEARNING SUMMARY

After studying this section you should be able to:

● *describe the composition of the Solar System*
● *explain how the orbits of satellites, planets and comets are due to gravitational forces*
● *explain why artificial satellites occupy different types of orbit according to their use.*

The Solar System

Apart from the nine known **planets**, there are a number of other objects contained within the Solar System:

● most of the planets have natural **satellites**, or **moons**

● there are artificial satellites in orbit around the Earth, and at times around other planets

> Asteroids range in size up to 1 km in diameter. Some asteroids have their own satellites.

● **asteroids** are small rocks that orbit the Sun. Most of the asteroids in the Solar System orbit between Mars and Jupiter

● there is an uncountable number of **comets**, with orbital times ranging from a few years to millions of years

● **meteors**, or shooting stars, come from the debris left behind in the orbit of a comet. When the Earth passes through this debris, the particles are pulled towards the Earth by its **gravitational field**; they become heated and glow as they pass through the atmosphere.

> Remember, forces act between objects. The size of the Earth's pull on the Moon is the same as the Moon's pull on the Earth.

An **attractive force** is required to keep a planet in orbit around the Sun and a moon in orbit around a planet. This force is the **gravitational force** that exists between all masses. The size of this force:

● depends on the **masses** of each of the objects involved; the greater their masses, the greater the force

● decreases with increasing distance between the objects; doubling the distance between two objects results in the force decreasing to one quarter of its value.

The diagram, **Fig. 6.1**, shows the orbits of the inner planets.

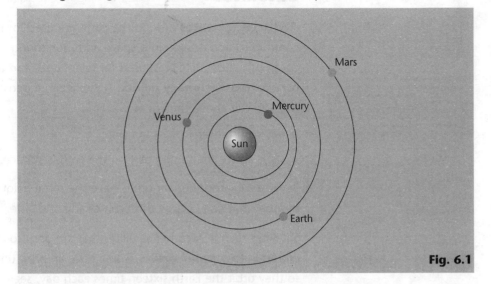

Fig. 6.1

With increasing distance from the Sun, the time to complete an orbit increases for two reasons:

- the orbital distance increases

- the speed of a planet in its orbit decreases as the gravitational force decreases.

> **Pluto's orbit is inclined at an angle of 17° to the planetary disc.**

Although planetary orbits are **elliptical**, all except those of Mercury and Pluto are nearly circular. The planets all orbit in the same direction around the Sun, and their orbits are contained within a disc.

Comets

Comets consist of dust and ices, frozen water and carbon dioxide. Their orbits are **highly elliptical**, they can orbit in any direction and in any plane. A comet is visible when it is close enough to the Sun for the **ices** to **vaporise** and is seen as a glowing ball of light. **Tails** appear on comets when they are as close to the Sun as the Earth is. The diagram, **Fig. 6.2**, shows the force on a comet at different parts of its orbit.

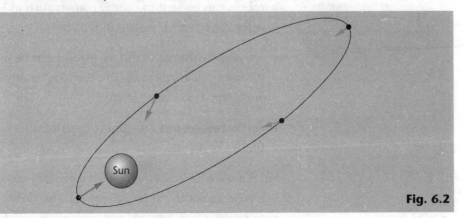

Fig. 6.2

> **The gravitational force on a comet acts partly in the direction of travel as it approaches the Sun, and partly in the direction opposite to travel as it moves away from the Sun.**

In this orbit:

- as it approaches the Sun the comet speeds up and the gravitational force becomes greater

- as it moves away from the Sun the comet slows down and the gravitational force becomes weaker.

Satellites

Artificial satellites around the Earth are used for navigation, surveillance, communication, seeing into space and monitoring the weather. Many communications satellites, used for telephones and television transmission, occupy **geostationary orbits**. A satellite in a geostationary orbit:

- is directly above the equator

- has an orbit time of 24 hours

- stays above the same point on the Earth's surface.

Because all geostationary orbits have the same height, there is a limit to the number that can occupy this orbit at any one time.

Satellites such as weather satellites that are used to monitor the Earth are placed in **low polar orbits**. These have an orbit time of one and a half hours, so they orbit the Earth sixteen times each day, seeing a slightly different view of the Earth on each orbit, as the Earth spins on its axis.

Is there life beyond the Solar System?

In astronomy, the billion is used in the American sense of meaning one thousand million.

Our Sun is one of the two hundred billion stars that make up the **Milky Way** galaxy, a collection of **stars** held together by gravitational forces. The Milky Way is a **spiral** galaxy, the stars forming the shape of a number of arms that spiral out from a central bulge. The galaxy is so vast that it takes light one hundred thousand years to pass between its extreme edges. It is rotating round at high speed, but the distances are so great that it takes two hundred million years for the Sun to complete one rotation.

There are more than a thousand billion known **galaxies** in the **Universe**. We know that life exists on one small star in our own galaxy. With so many galaxies and stars many people believe that the "chance" that created life on our planet must also have created life on other similar planets in the Universe. To try to detect such life:

The search for extra-terrestial intelligence is known as SETI.

- robots can be sent to nearby planets such as Mars to search for evidence of **microbes** or their fossilised remains

- the **atmospheres** of distant planets can be analysed, by detecting the different wavelengths of light that they transmit, to see if there is oxygen-enrichment due to plant life

- **radio telescopes** are used to search for radio signals from other advanced species of animal.

No evidence for any other life form existing has yet been found, but the immense size of the Solar System and the Universe means that direct exploration is very limited, and evidence in the form of radio waves could take many millions of years to reach us.

PROGRESS CHECK

1. Between which two planets is the asteroid belt?
2. The Sun's gravitational pull on Jupiter, the fifth planet out from the Sun, is greater than that on the Earth, the third planet. Suggest why.
3. What is the advantage of a weather satellite occupying a low polar orbit rather than a geostationary orbit?

1. Mars and Jupiter; 2. Jupiter is very massive; 3. In a low polar orbit, the satellite can monitor the whole of the Earth in one day.

6.2 Evolution

LEARNING SUMMARY

After studying this section you should be able to:

● *describe the life cycles of stars*
● *explain how movement of the galaxies supports the "Big Bang" theory*
● *explain how the future of the Universe depends on the amount of mass contained within it.*

The life of a star

Stars are born in clouds consisting of dust, hydrogen and helium. **Gravitational forces** cause regions of the cloud to **contract**, causing **heating**. As the core becomes hotter, the atoms lose their electrons and atomic nuclei collide at high speeds. Eventually a temperature is reached where hydrogen nuclei have enough energy to **fuse** together, resulting in the formation of helium nuclei. This in turn releases **energy**, and the star generates light and other forms of electromagnetic radiation from **nuclear fusion**.

The star enters its **main sequence**. In the main sequence:

● energy is released due to **fusion** of hydrogen nuclei in the core
● outward forces due to the high pressure in the core are balanced by gravitational forces.

> Nuclear fusion is the process in which nuclei join together. Very high speeds are needed for this to happen, due to the electrostatic repulsion between objects with similar charges.

Our Sun is a small star which is currently in its main sequence. This will end when there is not enough hydrogen left in the core to generate energy at the rate at which it is being radiated. When this happens:

● the Sun will cool and expand to become a **red giant**
● as the Sun expands the core will contract and become hot enough for the fusion of helium nuclei, resulting in the formation of the nuclei of carbon and oxygen
● the Sun will then contract, losing its outer layers and becoming a very hot, dense body called a **white dwarf**
● as energy is no longer being generated, the colour of a white dwarf changes as it cools and it eventually becomes an invisible **black dwarf**.

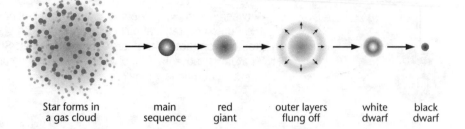

It takes millions of years for a white dwarf to cool and become a black dwarf.

Star forms in a gas cloud — main sequence — red giant — outer layers flung off — white dwarf — black dwarf

Fig. 6.3 The life cycle of a small star

More massive stars expand to become **red supergiants** after their main sequence. In a red supergiant:

- nuclear fusion in the contracting core results in the formation of nuclei of elements such as magnesium, silicon and iron

- the star is now generating energy again and becomes a **blue supergiant**

- when these nuclear reactions are finished the star cools and contracts again, glowing brightly as its temperature increases and becoming a **supernova**

- the supernova explodes, flinging off the outer layers to form a **dust cloud**

- the core that is left behind is a **neutron star**.

When astronomers notice that the brightness of a star is increasing, they realise that a spectacular event is about to happen when the supernova explodes.

Very dense neutron stars are called **black holes** because they are so dense that even light cannot escape from their gravitational fields. Black holes are detected by their effect on surrounding objects. They pull in gases from nearby stars. These gases reach very high speeds, and emit X-rays which can be detected as evidence for the existence of a black hole.

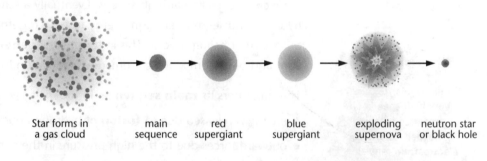

Star forms in a gas cloud — main sequence — red supergiant — blue supergiant — exploding supernova — neutron star or black hole

Fig. 6.4 The life cycle of a massive star

The existence of heavy elements in the inner planets and the Sun is evidence that our Solar System formed from the gas and dust flung off from the outer layers of an exploding supernova.

The past and future of the Universe

Evidence about the Universe's past history comes from examining the **spectra** of light given out by stars. When sources of waves such as light and sound move away from an observer, the waves detected by the observer have a longer wavelength than those emitted. This is known as **"red shift"**. When light from stars in other galaxies is analysed the results show that:

- light from almost all galaxies shows **red shift**

- the further away the galaxy, the greater the amount of red shift.

This suggests that the Universe is currently expanding. One theory that explains this expansion is the **"Big Bang"**. According to this model:

- the Universe started at a single point with an enormous explosion

- since the dawn of time, the Universe has been expanding and cooling

- the **microwave energy** that fills space is radiation left over from the explosion.

The "Big Bang" model is used to estimate the age of the Universe from measurements of the speeds of the galaxies. Current estimates show that they would have all been in the same place between 15 and 18 billion years ago.

The future of the Universe depends on the speeds at which the galaxies are moving apart and the amount of mass it contains. The three options are:

- there is enough mass for gravitational forces between the galaxies to stop the expansion and cause the Universe to contract, ending in a **"Big Crunch"**

- there is just enough mass to stop the expansion, leaving the Universe in a steady state

- there is insufficient mass to stop the expansion, and the Universe will continue to expand and cool.

> The apparent change in wavelength of sound waves is readily observed as an aircraft passes overhead or the siren of an emergency vehicle approaches and then recedes.

> The age of the Universe is only an estimate due to uncertainties in measuring the speeds of the galaxies and conjecturing how these speeds have changed over millions of years.

PROGRESS CHECK

1. What reaction occurs in the core of a star in its main sequence?
2. How does a black hole get its name?
3. What TWO pieces of evidence support the "Big Bang" theory of the origin of the Universe?

1. The fusion of hydrogen nuclei to form the nuclei of helium; 2. It cannot be seen because light cannot escape from it; 3. The red shift of light from other galaxies and the microwave radiation that fills space.

Sample IGCSE questions

1. There are thousands of artificial satellites in orbit around the Earth.

(a) Give TWO uses of artificial satellites. **[2]**

> *Artificial satellites are used for monitoring the weather ✓ and for television transmissions between the UK and the USA ✓.*

> Alternative acceptable answers include navigation, surveillance and for space telescopes.

(b) Some satellites orbit the Earth above the equator with an orbit time of 24 hours. These are called geostationary satellites. Explain why 24 hours is a suitable orbit time for some satellites. **[2]**

> *This is the time it takes for the Earth to rotate on its axis ✓, so the satellite is always above the same point on the Earth's surface ✓.*

> A common error at IGCSE is to state that a geostationary satellite moves at the same speed as the Earth – this is not the case.

(c) The graph shows how the orbit time of a satellite depends on its height above the Earth's surface.

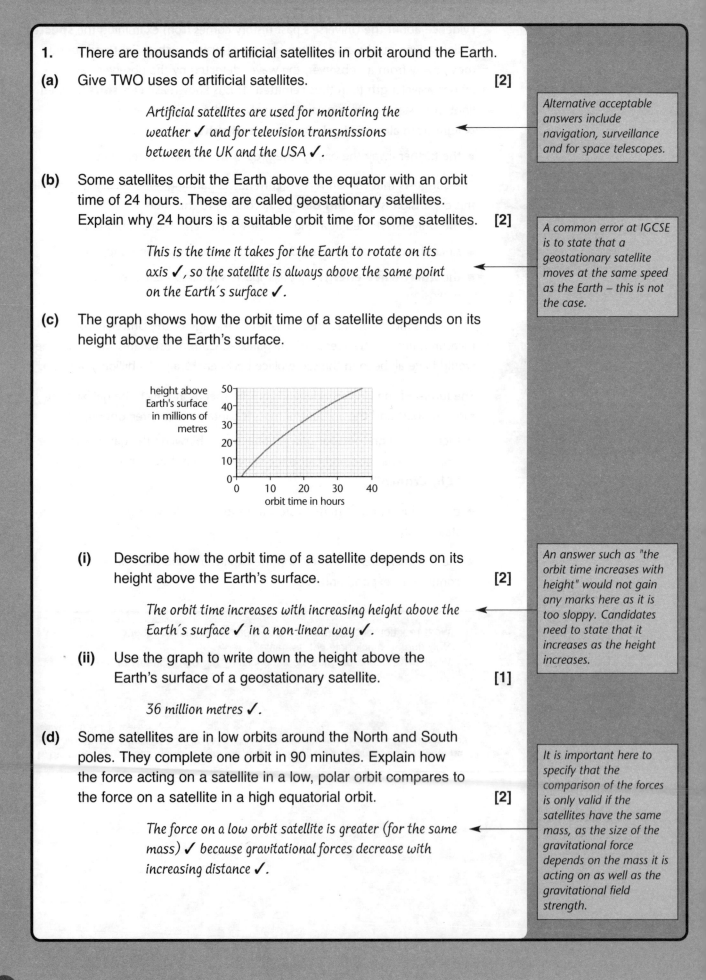

height above Earth's surface in millions of metres

orbit time in hours

(i) Describe how the orbit time of a satellite depends on its height above the Earth's surface. **[2]**

> *The orbit time increases with increasing height above the Earth's surface ✓ in a non-linear way ✓.*

> An answer such as "the orbit time increases with height" would not gain any marks here as it is too sloppy. Candidates need to state that it increases as the height increases.

(ii) Use the graph to write down the height above the Earth's surface of a geostationary satellite. **[1]**

> *36 million metres ✓.*

(d) Some satellites are in low orbits around the North and South poles. They complete one orbit in 90 minutes. Explain how the force acting on a satellite in a low, polar orbit compares to the force on a satellite in a high equatorial orbit. **[2]**

> *The force on a low orbit satellite is greater (for the same mass) ✓ because gravitational forces decrease with increasing distance ✓.*

> It is important here to specify that the comparison of the forces is only valid if the satellites have the same mass, as the size of the gravitational force depends on the mass it is acting on as well as the gravitational field strength.

Exam practice questions

1. The orbit time of a satellite depends on its height above the Earth's surface. A satellite used for television broadcasts orbits the Earth directly above the equator and has an orbit time of 24 hours.

   The diagram shows such an orbit.

   (a) Explain why this is called a geostationary orbit. [2]

   (b) Explain why satellites used for television broadcasts use geostationary orbits. [2]

   (c) Some communications satellites used for mobile telephones are in elliptical orbits. The diagram shows a satellite in an elliptical orbit.

   (i) Draw an arrow that shows the gravitational force on the satellite. [1]

   (ii) Describe how the force on the satellite changes as it approaches the Earth. [2]

   (iii) At which point on its orbit does the satellite have its greatest speed? [1]

2. Stars are formed within clouds consisting mainly of hydrogen, helium and dust.

   (a) Describe how a star is formed in such a cloud. [3]

   (b) Our Sun is a small star that also contains more massive elements. What does the presence of these elements suggest about the origin of the Sun?

   Give the reason for your answer. [2]

   (c) Our Sun is currently in its main sequence.

   (i) What happens to a star in its main sequence? [2]

   (ii) What is likely to happen to the Sun at the end of its main sequence? [3]

Exam practice questions

3.

(a) What evidence is there that the galaxies are moving away from each other? [3]

(b) The Andromeda galaxy is moving towards our galaxy, the Milky Way.
How does light detected from the Andromeda galaxy differ from that detected from other galaxies? [2]

(c) **(i)** What additional evidence is there to support the "Big Bang" theory? [1]

(ii) How does this theory picture the evolution of the Universe up to the present time? [3]

(d) One possible future of the Universe is the "Big Crunch".
What has to happen to cause this? [3]

4.

(a) In 1610 Galileo trained his newly-discovered telescope on Jupiter and discovered four star-like objects.

(i) Suggest why no-one had reported seeing these before. [1]

After observing their position for several days, he concluded that they were moons of Jupiter.

(ii) What could he have observed to lead to this conclusion? [2]

(b) The table gives some data about the Galilean moons.

Moon	Orbital distance in m	Orbital period in s	Radius in m	Density in g/cm^3
Io	4.3×10^8	1.5×10^5	1.8×10^6	3.6
Europa	6.7×10^8	3.1×10^5	1.6×10^6	3.0
Ganymede	1.1×10^9	6.2×10^5	2.6×10^6	1.9
Callisto	1.9×10^9	1.4×10^6	2.4×10^6	1.8

(i) What is the relationship between orbital period and orbital distance? [1]

(ii) What evidence in the table indicates that the two outer moons are icy while the two inner moons are rocky? [1]

(iii) In what other way are the two outer moons different to the two inner moons? [1]

(iv) One feature of the Galilean moons is similar to that of the Earth's moon; they each have the same face pointing towards Jupiter at all times.

What is the time period of rotation of Ganymede on its own axis? [1]

Chapter 1 General physics

1 (a) (i) B is the forwards push [1] of the wheel on the road. [1]

    (ii) C [1]

    (iii) Wet leaves reduce the friction between the wheel and the road [1] which would cause the wheel to slip. [1]

(b) (i) Air resistance. [1]

    (ii) 60 N [1] forwards. [1]

    (iii) Acceleration = force ÷ mass [1]

$$= 60 \text{ N} \div 90 \text{ kg } [1]$$
$$= 0.67 \text{ m/s}^2 \text{ } [1]$$

    (iv) The resistive force increases [1] The unbalanced force decreases [1] causing the acceleration to decrease. [1]

2 (a) (i) Thinking distance is proportional to speed. [1]

    (ii) As speed increases so does braking distance, but not in direct proportion. [1]

(b) (i) 16 m [1]

    (ii) 21 m [1]

    (iii) 37 m [1]

(c) (i) Any two from: drugs, alcohol, tiredness, driver's concentration. [2]

    (ii) Any two from: vehicle mass, condition of brakes, condition of road surface. [2]

3 (a) (i) Acceleration = change in velocity ÷ time taken [1]

$$= 30 \text{ m/s} \div 3 \text{ s } [1]$$
$$= 10 \text{ m/s}^2 \text{ } [1]$$

    (ii) Downwards. [1] The negative gradient of the graph shows that the acceleration is in the opposite direction to the upward velocity. [1]

    (iii) Force = mass × acceleration [1]

$$= 0.020 \text{ kg} \times 10 \text{ m/s}^2 \text{ } [1]$$
$$= 0.20 \text{ N } [1]$$

(b) (i) 1.5 s [1]

This is when the sign of the velocity changes. [1]

    (ii) Distance travelled = average speed × time. [1]

$$= \tfrac{1}{2} \times 15 \text{ m/s} \times 1.5 \text{ s } [1]$$
$$= 11.25 \text{ m } [1]$$

4 (a) Change in momentum = mass × change in velocity [1]

$$= 13\,000 \text{ kg} \times 2\,500 \text{ m/s } [1] = 3.25 \times 10^7 \text{ kg m/s. } [1]$$

(b) Change in speed = change in momentum ÷ mass [1]

$$= 3.25 \times 10^7 \text{ kg m/s} \div 2.7 \times 10^6 \text{ kg } [1] = 12.0 \text{ m/s. } [1]$$

(c) The mass of the rocket decreases [1] so for the same increase in momentum each second there is a greater increase in speed [1].

5 (a) 150 N [1] from right to left [1].

(b) The total momentum remains zero [1]. The larger boy gains momentum directed towards the right [1] so the smaller boy gains an equal amount of momentum in the opposite direction [1].

(c) 30 kg × v = 50 kg × 1.5 m/s [1], v = 50 kg × 1.5 m/s ÷ 30 kg [1] = 2.5 m/s [1].

6 The completed vector diagram is [1]:

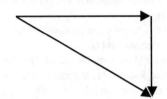

The speed of the ball is 18.0 m/s [1] at an angle of 56° to the vertical [1].

7 (a) The path is curved, with the slope of the curve increasing [1]. This is because the dart travels equal distances in equal time intervals horizontally [1] but increasing distances in successive equal time intervals vertically [1].

(b) time = distance ÷ speed [1] = 1.8 m ÷ 12.4 m/s [1]
$$= 0.145 \text{ s } [1]$$

(c) $s = ut + \tfrac{1}{2}at^2$ [1] = $\tfrac{1}{2} \times 10$ m/s$^2 \times (0.145$ s$)^2$ [1]
$$= 0.11 \text{ m } [1].$$

8 (a) A satellite in a circular orbit is accelerating towards the Earth [1]. An unbalanced force is needed to cause this acceleration [1].

(b) The gravitational pull [1] of the Earth on the satellite [1].

(c) The mass of the satellite [1], its speed [1] and radius of orbit [1].

(d) The polar satellite has a greater speed [1] and a shorter orbital distance [1].

(e) (i) $T = 2\pi r/v$ [1] = $2 \times \pi \times 7.1 \times 10^6$ m ÷ 7.5×10^3 m/s [1]
$$= 5.9 \times 10^3 \text{ s } [1].$$

    (ii) $a = v^2/r$ [1] = $(7.5 \times 10^3$ m/s$)^2 \div 7.1 \times 10^6$ m [1]
$$= 7.9 \text{ m/s}^2 \text{ } [1].$$

    (iii) $F = m \times a$ [1] = 1.2×10^3 kg × 7.9 m/s^2 [1]
$$= 9.5 \times 10^3 \text{ N } [1].$$

    (iv) Towards the centre of the Earth [1].

9 (a) Change in momentum = mass × change in speed [1]
$$= 0.046 \text{ kg} \times 60 \text{ m/s } [1] = 2.76 \text{ kg m/s. } [1].$$

(b) Force = change in momentum ÷ time [1] = 2.76 kg m/s ÷ 4×10^{-4} s [1] = 6.9×10^3 N [1].

(c) The mass of the club is large compared to that of the ball [1] so although its momentum changes by the same amount it hardly affects the speed [1].

Chapter 2 Thermal physics

1 (a) B [2]

(b) D [1]

(c) B [1]

(d) BC and DE are flat because, although we are supplying heat, the temperature is not rising. [1]

The heat energy supplied goes into changing the state of the substance. [1]

BC – the latent heat of fusion as ice changes to water. [1]

DE represents the latent heat of vaporization as water changes to steam. [1]

This latent heat of fusion has a very high value as it involves complete breaking during the change from the liquid to the gaseous phase. [1]

Exam practice answers

2 (a) 27°C = 300K [1]

(b) 1.2 x 500=P_{new} x 200, so that new pressure = 3 N/m² [1]

(c) Boiling point of water = 273K [1]

(d) As the gas is heated, energy is given to its molecules [1] this makes them move more quickly [1] the number of collisions made by the molecules with the piston each second increases [1] each collision is more energetic than that at a lower temperature [1] so the piston is pushed backwards and the gas expands to a new volume. [1]

3 (a) The pressure is due to collisions [1] between the particles and the walls of the cylinder [1].

(b) The average speed of the particles increases [1] so the collisions are more frequent [1] and more force is exerted (on average) at each collision [1].

(c) $P_1/T_1 = P_2/T_2$ so $P_2 = P_1 \times T_2 \div T_1$ [1] = 3.5×10^5 Pa × 323 K ÷ 293 K [1] = 3.9×10^5 Pa [1].

(d) The cylinder would absorb energy from the Sun's radiation [1] which would raise the temperature of the contents [1] and could cause the cylinder to explode due to the increased pressure [1].

4 (a) $E = m \times c \times \Delta\theta$ [1] = 1.50 kg × 4.2×10^3 J/kg °C × 90°C [1] = 5.67×10^5 J [1].

(b) Energy is needed to heat the element [1] and the casing [1] and some energy is lost to the surroundings [1].

Chapter 3 Properties of waves

1 (a) Any two from: infra-red, light, radio, ultraviolet, gamma. [2]

(b) Ultraviolet. [1]

(c) Infra-red. [1]

(d) Sound/ultrasound. [1]

(e) (i) The device detects the echo. [1]
   The time is measured. [1]
   This is halved and multiplied by the speed of sound. [1]

   (ii) Furniture would scatter the sound. [1] Giving multiple reflections. [1]

2 (a) Diffraction. [1]

(b) The width of the gap [1] and the wavelength. [1]

(c) The width of the doorway is approximately one wavelength for the sound wave [1] but is many wavelengths for light. [1]

3 (a) Ray to centre of lens continued straight [1]. Ray parallel to principal axis drawn through principal focus [1]. Upright arrow drawn where these cross [1].

(b) 13.3 cm [1].

(c) Any two from: object is inverted, image is upright; image is further away from lens than object; image is larger than object [2].

(d) Projector or enlarger [1].

(e) So that the image is seen the correct way up on the screen [1].

4 (a) The frequency of vibration when an object is displaced and allowed to vibrate freely [1].

(b) When an object is forced to vibrate at its natural frequency [1] large amplitude vibrations build up [1].

(c) (i) As the length of the wire increases [1] the natural frequency decreases [1].

   (ii) 260 Hz [1].

   (iii) 1.6 m [1].

   (iv) $v = f \times \lambda$ [1] = 260 Hz × 1.6 m [1] = 416 m/s [1].

5 (a) Diffraction [1].

(b) A patch of light [1] the beams do not overlap [1].

(c) Bright and dark bands [1]. Bright bands occur due to constructive interference [1]. The dark bands are due to destructive interference [1].

(d) The difference in path length between two waves arriving at the same place [1].

(e) A whole number of wavelengths or an even number of half wavelengths [1].

6 (a) Digital [1].

(b) Analogue [1].

(c) (i) Magnetic [1].

   (ii) The magnetic domains are realigned [1] from random to an ordered arrangement [1].

   (iii) As the tape moves, there is a changing magnetic field in the core [1]. This causes an alternating voltage in the coil [1] by electromagnetic induction [1].

7 (a) There is less spreading of the beam [1] so less energy is lost [1].

(b) Space waves [1] as the others cannot travel through the Earth's atmosphere [1].

(c) (i) Any two from: it takes 24 hours to orbit the Earth; the satellite remains above the same point on the Earth's surface; the orbit is equatorial [2].

   (ii) The receiving aerial can be fixed [1].

(d) It receives waves [1], amplifies them [1] and retransmits them [1].

(e) Digital signals can carry more information [1]. They can be restored to their original condition [1] giving better quality reception [1].

(f) Short wavelength waves have a high frequency and can carry a lot of information [1]. Short wavelength waves can travel through the ionosphere [1]. There is less diffraction at the dishes with short wavelength waves [1].

Chapter 4 Electricity and magnetism

1 (a) 1.6 V [1]

(b) resistance = voltage ÷ current [1]
   = 1.6 V ÷ 0.5 A [1]
   = 3.2 Ω [1]

(c) A filament lamp. [1]
   In a filament lamp the resistance increases as the filament gets hotter. [1]

(d) The resistance increases. [1]
   The resistance at 6 V is 4 Ω. (or equivalent calculation) [1]

2 (a) (i) Electron. [1]

   (ii) Negative. [1]

(b) Electrons move [1] from the ground [1] to neutralise the charge on the airframe. [1]

(c) The aircraft can discharge when it lands [1] by charge passing [1] between the airframe and the ground [1]

3 (a) Current = charge ÷ time [1]
   = 300 C ÷ 60 s [1]
   = 5 A [1]

(b) energy = charge × voltage [1]
   = 300 C × 12 V [1]
   = 3600 J [1]

(c) power = current × voltage [1]
   = 5 A × 12 V [1]
   = 60 W [1]

4 (a) current = power ÷ voltage [1]
   = 8400 W ÷ 240 V [1]
   = 35 A [1]

(b) The large current causes heating in the cables. **[1]** Thick cables have a low resistance so little heating occurs. **[1]**

(c) It acts faster than a fuse **[1]**

It is easily reset **[1]**

(d) energy transfer = 8.4 kW × 3.5 h **[1]**

= 29.4 kWh **[1]**

cost = 29.4 × 7p = 206p **[1]**

5 (a) A **[1]**

(b) B **[1]**

(c) C **[1]**

(d) C **[1]**

(e) power = current × voltage **[1]**

= 2.5 A × 12.0 V **[1]**

= 30 W **[1]**

6 (a) Electromagnetic induction. **[1]**

(b) (i) Moving the magnet. **[1]** Moving the coil. **[1]**

(ii) Any three from:

the number of turns on the coil, the area of the coil, the speed of movement, the strength of the magnetic field. **[3]**

(iii) Reverse the direction of movement. **[1]** Reverse the poles of the magnet. **[1]**

(c) (i) The current from the battery is d.c. **[1]** This creates a constant magnetic field. **[1]**

(ii) The current in the input coil is continually changing. **[1]** This produces a changing magnetic field in the output coil. **[1]**

(iii) The ratio of the voltages is 1:4. **[1]** The numbers of turns are in the same ratio, so there are 800 on the output coil. **[1]**

7 (a) Current in the conductors causes heating. **[1]** The water removes excess heat. **[1]**

(b) (i) They are much cheaper to install. **[1]** They are cheaper to run as the air acts as a coolant. **[1]**

(ii) There is not enough space for pylons. **[1]** Overhead conductors could be blown against buildings in windy conditions. **[1]**

(c) Any three from:

The power of the Sun's radiation is too low to make solar heating or solar cells economic.

There are few fast-flowing rivers and streams to generate hydroelectricity.

Rocks below the ground are not hot enough to generate steam to drive steam turbines.

There is not enough land where wind blows all the time to generate electricity in significant quantities. **[3]**

8 (a) A step-up transformer has more secondary turns than primary turns. **[1]** A step-down transformer has fewer secondary turns than primary turns. **[1]**

(b) Transformers are used to step up the voltage before electricity is transmitted, **[1]** and to step down the voltage before electricity is supplied to consumers. **[1]**

(c) The energy lost due to heating of the wires is proportional to the square of the current. **[1]** Distributing electricity at high voltages enables small currents to be used. **[1]**

(d) The voltage of a direct current cannot be changed. **[1]** Transformers allow alternating voltages to be stepped up and stepped down. **[1]**

9 (a) Any two from: Sun, wind, hydroelectric, tidal, geothermal, biomass. **[2]**

(b) Any two from: gas, coal, oil. **[2]**

(c) (i) Gas-fired power stations are more efficient than coal-fired power stations. **[1]**

(ii) Gas is burnt in a combustion chamber and the

pollutants produced inside can be disposed off safely. **[2]**

(iii) Coal tar residue is not produced as gas is not a solid fuel. **[3]**

(e) (i) Advantage: it does not use fossil fuels/does not cause atmospheric pollution. **[1]**

Disadvantages: any two from: unsightly, noisy, a large area of land is needed. **[2]**

(ii) There are few fast-flowing rivers and streams. **[1]** These are mainly in Wales and Scotland. **[1]** Obtaining energy from the tides is costly and unreliable. **[1]**

(iii) Advantages: it does not use up reserves of fossil fuel, **[1]** it does not release carbon dioxide into the atmosphere. **[1]**

Disadvantages: any two from: leaks can cause pollution of the atmosphere and local rivers and streams, it is very expensive to close down a nuclear power station safely, there is a constant problem of disposing of nuclear waste. **[2]**

10 (a) Arrow drawn in the direction N to S. **[1]**

(b) (i) These sides are carrying a current at right angles to the direction of the magnetic field. **[1]**

(ii) These sides are carrying a current parallel to the direction of the magnetic field. **[1]**

(c) (i) Each force has a turning effect about the pivot (the central axis). **[1]** The moments both act in a clockwise direction. **[1]**

(ii) Reverse the current. **[1]** Reverse the direction of the magnetic field. **[1]**

(iii) Increase the voltage/current. **[1]** Increase the strength of the magnetic field. **[1]**

11 (a) Logic gate 1 is OR **[1]**. Logic gate 2 is AND **[1]**.

(b) The completed table is:

input A	input B	input C	output
0	0	0	0
1	0	0	0
0	1	0	0
1	1	0	0
0	0	1	0
1	0	1	1
0	1	1	1
1	1	1	1

Award one mark for each two rows correct **[4]**.

(c) C **[1]**.

(d) The alarm is on **[1]** as the OR gate gives a 1 output when both inputs are 1 **[1]**.

12 (a) The resistance of the thermistor increases **[1]**. The voltage across the thermistor goes up and that across the variable resistor goes down **[1]**. The heater is switched on when the input to the NOT gate is below the threshold for it to be high (a 0) **[1]**.

(b) The temperature at which the heater is switched on can be adjusted **[1]**.

(c) The voltage across the thermistor = 3.0 V **[1]**. Resistance of variable resistor = 2/3 × 45 Ω **[1]** = 30 Ω **[1]**.

13 (a) (i) The logic gate **[1]**.

(ii) The alarm **[1]**.

(b) The completed tables are:

door switch	
input conditions	output
door open	1
door closed	0

temperature sensor	
input conditions	output
less than 10°C	0
more than 10°C	1

One mark for each table completed correctly [2].
(c) (i) When the door is closed [1] or the temperature is below 10°C [1].
   (ii) Place a NOT gate [1] between the NOR gate and the alarm OR between each input and the NOR gate [1].

14 (a) The resistance of the LDR increases [1]. The output voltage increases [1].
   (b) The voltage across the fixed resistor = 2.0 V [1]. The resistance of the fixed resistor = 2/3 × 2000 Ω [1] = 1333 Ω [1].
   (c) The fixed resistor has a greater share of the voltage [1] so the output voltage decreases [1].

15 (a) 3 Ω [1].
   (b) The completed table is

Resistor	Current in A	Voltage in V
2 Ω	2.4A	4.8
6 Ω	1.2A	7.2
8 Ω	0.9A	7.2
24 Ω	0.3A	7.2

1 mark for each row completed correctly [4].

16 (a) Thermionic emission [1].
   (b) The electrons need to gain energy to leave the metal [1], when the filament is heated the electrons gain energy [1].
   (c) They are attracted to the positive anode [1] and repelled from the negative cathode [1].
   (d) $E_k = e \times V = 1.6 \times 10^{-19}$ C × 2 500 V [1] $= 4.0 \times 10^{-16}$ J [1].

17 (a) The trace shows both positive and negative voltages [1].
   (b) (i) +6 V [1].
      (ii) −6 V [1].
   (c) Time for one cycle = 8×10^{-3} s. [1]. $f = 1/T = 1 \div 8 \times 10^{-3}$ s [1] = 125 Hz [1].

Chapter 5 Atomic and particle physics

1 (a) Beta particles are negatively charged. [1] They consist of high-speed electrons. [1] They are absorbed by a few mm of aluminium or other metal. [1]
   Gamma rays are not charged. [1] They are short-wavelength electromagnetic radiation. [1] Their intensity is reduced by thick lead or concrete. [1]
   (b) Here is a completed graph.

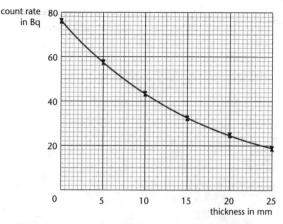

Marks are awarded for: choice of scale and correct labelling of axes [1] correct plotting of points [2] drawing of smooth curve. [1]
   (ii) 12.5 mm. [1]
      By using the graph to find the thickness required to halve the count rate from 76 Bq to 38 Bq. [1]
   (iii) 37.5 mm. [1]
      15 Bq is 1/8 = 1/2 × 1/2 × 1/2 of 120 Bq. [1]
      So the thickness of lead needs to halve the intensity three times. [1]

2 (a) Isotopes of an element have the same number of protons [1] but different numbers of neutrons. [1]
   Half-life is the average time it takes [1] for the number of undecayed nuclei to halve. [1]
   (b) (i) So that the radiation penetrates the flesh [1] and can be detected by the camera. [1]
      (ii) Alpha and beta radiation would be absorbed by flesh [1] and could cause damage to cells. [1]
   (c) 6 hours is long enough for the material to be active enough to be detected when it has circulated. [1] And short enough to minimise the risk of damaging the patient by exposure to radiation. [1]

3 (a) (i) New supplies of carbon-14 are constantly being absorbed [1] to replace that lost due to decay. [1]
      (ii) As no new carbon-14 is being absorbed, the number of undecayed nuclei decreases. [1] So the number decaying each second decreases, as this is proportional to the number of undecayed nuclei present. [1]
   (b) (i) 500 Bq. [1]
      (ii) Time taken for rate of decay to halve = 5 600 years. [1] Indicated on graph. [1]
      (iii) If the sample were 0.5 kg the rate of decay would be 60 Bq. [1] This gives an age of 6 600 years. [1]

4 (a) Any two from: sodium chloride is soluble in water, gamma radiation can be detected on the surface, the half-life is long enough for water to flow through the pipes and the leak to be detected. [2]
   (b) Radioactive sodium chloride is dissolved in the water at the input. [1] When water has flowed all the way along the pipes, a detector is used to find the leak. [1]
   (c) (i) A Geiger-Müller tube. [1]
      (ii) The reading would go up. [1] Since radioactive water accumulates around the leak. [1]
   (d) Three half-lives [1] = 45 hours. [1]

5 (a) Gamma radiation is the emission of energy, not particles, and has no effect on A or Z values. [1].
   (b) Strontium-90 lies above the stability line [1] so it emits

β⁻ radiation **[1]**

(c) $_{38}^{90}\text{Sr} \rightarrow \,_{39}^{90}\text{Y} + \,_{-1}^{0}\text{e}$ 1 mark for each correct symbol on the right hand side of the equation **[2]**

(d) (i) For the least massive stable nuclei the neutron-proton ratio is 1 **[1]**. This increases as the nuclear mass increases **[1]** to a value of about 1.4 for the most massive nuclei **[1]**.

  (ii) For these nuclei to be stable the neutron-proton ratio is 1 **[1]**, this would hardly change by emission of an alpha particle **[1]**.

Chapter 6 The Earth and beyond

1 (a) The satellite's orbit time is the same as the time it takes the Earth to rotate on its axis. **[1]** So the satellite remains above the same point on the Earth's surface. **[1]**

(b) Fixed aerials that point towards the satellite are used to receive the transmissions. **[1]** So the satellite needs to remain in a fixed position relative to the aerials. **[1]**

(c) (i) Arrow pointing towards the centre of the Earth. **[1]**

  (ii) The size of the force increases. **[1]** The direction of the force changes so that it is always towards the centre of the Earth. **[1]**

  (iii) When it is closest to the Earth. **[1]**

2 (a) Gravitational forces cause parts of the star to contract. **[1]** This causes heating. **[1]** A star is created when it is hot enough for fusion reactions to occur. **[1]**

(b) It formed from the remains of an exploding supernova. **[1]** Since these elements must have formed in a star after the main sequence. **[1]**

(c) (i) Hydrogen nuclei fuse together **[1]** to form the nuclei of helium. **[1]**

  (ii) It will expand to become a red giant. **[1]** The outer layers will be flung off. **[1]** The core will then contract to become a white dwarf. **[1]**

3 (a) The wavelength of light received from other galaxies is lengthened **[1]** and shifted towards the red end of the spectrum. **[1]** This happens when objects are moving away from each other. **[1]**

(b) The wavelength of light received from Andromeda is shortened. **[1]** It shows "blue shift". **[1]**

(c) (i) Microwave radiation fills space. **[1]**

  (ii) The Universe started with an enormous explosion. **[1]** Since then it has been expanding and cooling. **[1]** Stars and galaxies have formed from clouds of dust and gas. **[1]**

(d) Gravitational forces need to be big enough to stop the expansion **[1]** and cause the Universe to contract. **[1]** This will only happen if there is sufficient mass in the Universe. **[1]**

4 (a) (i) They cannot be seen with the naked eye. **[1]**

  (ii) The movement of the moons **[1]** around Jupiter. **[1]**

(b) (i) The greater the orbital distance, the greater the period. **[1]**

  (ii) The two inner moons are denser than the outer ones. **[1]**

  (iii) They are larger. **[1]**

  (iv) 6.2×10^5 s. **[1]**

Formulae

This list shows the relationships that awarding bodies are not allowed to provide for candidates.

speed = $\dfrac{\text{distance}}{\text{time}}$

acceleration = $\dfrac{\text{change in velocity}}{\text{time taken}}$ $a = \dfrac{v-u}{t}$

force = mass × acceleration $F = m \times a$

density = $\dfrac{\text{mass}}{\text{volume}}$

work done = force × distance moved in direction of force $W = F \times x$

energy transferred = work done

kinetic energy = ½ × mass × speed² $KE = \frac{1}{2} \times m \times v^2$

change in potential energy = mass × gravitational field strength × change in height $GPE = m \times g \times h$

weight = mass × gravitational field strength $W = m \times g$

pressure = $\dfrac{\text{force}}{\text{area}}$ $P = \dfrac{F}{A}$

moment = force × perpendicular distance to pivot

charge = current × time $Q = I \times t$

voltage = current × resistance $V = I \times R$

electrical power = voltage × current $P = I \times V$

wave speed = frequency × wavelength $v = f \times \lambda$

$\dfrac{\text{voltage across secondary}}{\text{voltage across primary}} = \dfrac{\text{number of turns on secondary}}{\text{number of turns on primary}}$ $\dfrac{V_s}{V_p} = \dfrac{n_s}{n_p}$

energy transferred = power × time $E = P \times t$

cost of electricity = number of Units × cost per Unit

efficiency = $\dfrac{\text{useful energy transferred by device}}{\text{total energy supplied to device}}$

energy transferred = voltage × charge $E = V \times Q$

the equations of motion $v = u + at$
$v^2 = u^2 + 2as$
$s = ut + \frac{1}{2}at^2$

momentum = mass × velocity $p = m \times v$

energy transfer = mass × specific heat capacity × temperature change $E = m \times c \times \Delta\theta$

efficiency = $\dfrac{\text{useful work or energy output}}{\text{total energy input}}$

Model Test Paper I
PHYSICS

M.M: 90   **Time:** 1 hour 15 min

1 **(a)** What is meant by the term 'renewable sources of energy'?

.... **[1]**

....

....

(b) Name any three renewable sources of energy.

.... **[3]**

....

....

(c) Name one advantage and one disadvantage of any renewable source of energy.

Advantage: .... **[1]**

....

Disadvantage: .... **[1]**

....

(d) State the energy changes involved in the following:

A cyclist pedals up a steep hill. When he reaches the top; he freewheels down to the bottom of the hill, where he applies the brakes and comes to a halt.

(i) Pedalling uphill:

.... changes to .... **[1]**

(ii) Freewheeling downhill:

.... changes to .... **[1]**

(iii) Applying the brakes:

.... changes to .... **[1]**

(e) State the process in which energy is transferred for each of the following:

(i) the heat and light from the Sun reach the Earth: .... **[1]**

(ii) a radiator heats up a room. .... **[1]**

(Total : 11)

2 The diagram below shows a 62 kg swimmer standing still on the edge of a diving board.

(a) One of the forces acting on the swimmer is her weight. What other force acts on the swimmer as she stands still at the edge of the diving board?

.... **[1]**

(b) Calculate the weight W of the swimmer.
The gravitational field strength $g = 10$ N/kg.

W = ... N **[2]**

(c) (i) The weight W of the swimmer exerts a moment about the point **X**.
State whether this moment is clockwise or anticlockwise.

.... **[1]**

(ii) The diving board is resting on the support **Y**. Discuss whether the force F at this support is greater than or less than the weight of the swimmer.

....

.... **[2]**

(d) The swimmer gently walks over the edge of the diving board and drops vertically down into the pool water below. The velocity against time graph below shows the motion of the swimmer in free fall and in the water.

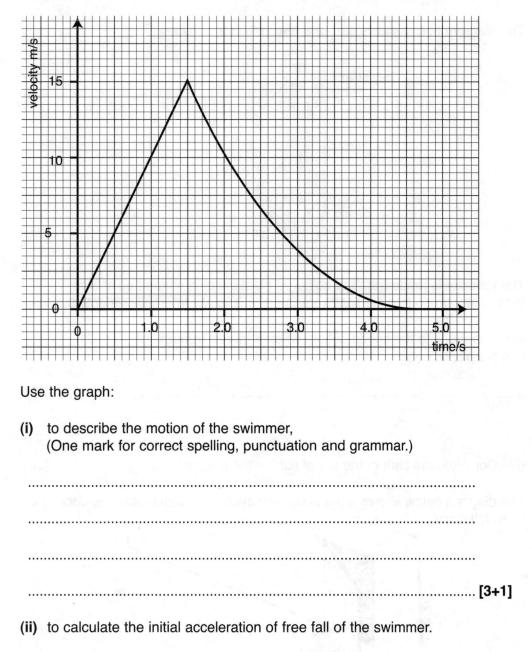

Use the graph:

(i) to describe the motion of the swimmer,
(One mark for correct spelling, punctuation and grammar.)

....

....

....

.... **[3+1]**

(ii) to calculate the initial acceleration of free fall of the swimmer.

acceleration = ... m/s^2 **[3]**

[Total: 13]

3 (a) Visible light is a transverse wave. Explain what is meant by a transverse wave.

....

.... **[1]**

(b) The diagram below shows part of a rear reflector of a car.

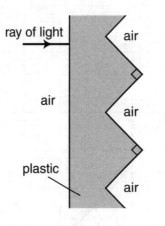

ray of light ——→

air

air

air

air

plastic

The reflector is made from red coloured plastic that is shaped as shown in the diagram so that the light from the rear is totally internally reflected.

(i) What happens to the speed of light when it travels from the air into the plastic?

.... **[1]**

(ii) Complete the path of the ray of light in the diagram. **[2]**

(c) The diagram below shows a piano being played in a room where the door has been left open.

Suggest why a person on the other side of the open door can hear the piano clearly even though he cannot directly see it.

....

....

.... **[3]**

(d) The diagram below shows two loudspeakers connected to the same a.c. supply.

a.c. supply
220 Hz

loudspeakers

person listening
to sound from
loudspeakers

(i) With the switch **A** closed, and switch **B** open, a person at point **X** hears a very loud sound. The frequency of the sound is 220 Hz. Calculate the wavelength of the sound given the speed of sound in air is 340 m/s.

wavelength = ... m   **[3]**

(ii) When both switches **A** and **B** are closed, the person suddenly hears no sound even though each loudspeaker is still emitting sound. Explain why this happens.

....

....

.... **[2]**

[Total: 12]

4 **(a)** Complete the ray diagram below to show what is meant by the
focal length of a diverging (concave) lens. **[2]**

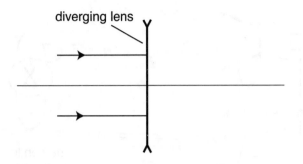

(b) The diagram below shows a converging (convex) lens used as
a magnifying glass.

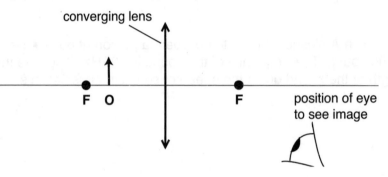

The point marked **F** is the principal focus of the converging lens.
The object **O** lies between the principal focus and the centre of the lens.
The image produced by the lens is virtual.

(i) Explain what is meant by a *virtual image*.

....

.... **[1]**

(ii) Draw a ray diagram to locate the position of the virtual image formed
by the lens. Label the image **I**. **[3]**

(iii) Apart from being a virtual image of the object **O**, state two other
properties of this image.

....

.... **[2]**

[Total: 8]

5 A 24V battery is connected across an 8Ω resistor in series with a parallel
combination of a 12Ω resistor and a 6Ω resistor as shown in the diagram.

24V

12Ω

8Ω

6Ω

(a) Calculate:

(i) the voltage across the 8Ω resistor

....**[3]**

(ii) the current through the 6Ω resistor.

....**[3]**

(b) What is the unit of electricity energy used, when calculating domestic usage?

....**[2]**

(c) The power ratings of some electrical appliances are given in the following table:

DVD Player	20W
Lightbulb	100W
Television	100W
Refrigerator	150W

If a person uses all the above devices for 10 hours per day in December and if the cost of one unit is $0.10, calculate:

(i) the total units consumed

....**[3]**

(ii) the total money due.

.... **[3]**

[Total: 14]

6 **(a)** The diagram below shows two coils made from insulated copper wires wrapped round a soft-iron rod.

The coil **A** is connected in series with a switch **S**, an ammeter and a battery. The coil **B** is connected to a voltmeter.

The switch **S** is closed. The voltmeter connected to coil **B** shows a deflection and after a short period of time it shows no reading even though there is constant current shown by the ammeter. Explain these observations.
(One mark for a correctly ordered answer.)

....

....

....

....

....

....

.... **[3+1]**

(b) A mobile phone charger unit has a transformer that steps-down the voltage from 230 V to 3.8 V. The primary coil has 5200 turns and an input current of 12 mA. The transformer is 100% efficient.

Calculate:

(i) the number of turns on the secondary coil,

turns = ... **[3]**

(ii) the current in the secondary coil.

current = ... A **[3]**

(c) Electric power is distributed by the National Grid at a very high voltage of 400 000 V. Suggest why this voltage is so high.

....

.... **[1]**

[Total: 11]

7 Most of the carbon dioxide in the Earth's atmosphere contains atoms of the stable isotope carbon-12. A small percentage of the carbon dioxide also contains radioactive atoms of the isotope carbon-14.

(a) Explain what is meant by an *isotope*.

....

.... **[2]**

(b) Explain what is meant by background radiation. Name one source for this radiation.

....

.... **[2]**

(c) Carbon dioxide is absorbed by all living trees. The ratio of carbon-12 atoms to carbon-14 atoms in all living trees is a constant. When a tree dies it stops absorbing carbon-14 from the atmosphere. The carbon-14 already in the tree starts to decay. The half-life of carbon-14 atoms is 5600 years.

(i) An atom of carbon-14 may be represented as $^{14}_{6}$ C. In the nucleus of carbon-14, how many protons and neutrons are there?

protons: …………

neutrons: …………   **[2]**

ii) An archaeologist discovers a wooden spear. A sample of 1.0 gram of carbon taken from the wooden spear gives an average of 22.5 counts per hour and a 1.0 gram sample of carbon taken from a living tree gives an average of 90 counts per hour. Assuming that the ratio of carbon-12 to carbon-14 atoms has remained constant since the spear was made, determine the age of the wooden spear in years.

age = …………. years **[3]**

[Total: 9]

8 The diagram shows a uniform steel girder, of length 8 metres and mass 120 kg supported at two points P and Q.

A man of mass 80 kg stands on the girder, 2 m from P.

(a) Draw arrows on the diagram to show the weight of the man and of the girder, and the upward forces at P and Q.

………………………………………………………………………………………………… **[2]**

(b) By taking moments about P and then about Q, find the force at each of the supports.

…………………………………………………………………………………………………

…………………………………………………………………………………………………

………………………………………………………………………………………………… **[2]**

(c) The man now walks 5 metres towards Q. What are the forces at each of the supports now? (Assume the value of acceleration due to gravity, g, as 10 N/kg)

…………………………………………………………………………………………………

…………………………………………………………………………………………………

.... **[2]**

(d) The soles of the man's shoes have a surface area of 80cm². Calculate the pressure he exerts on the girder:

(i) in Ncm⁻² **[1]**

(i) in Pa. **[1]**

(e) The girder spans a pond 3 metres below it. If the man throws a stone into the pond, it will accelerate as it falls through the air, then decelerate when it enters the pond, and then fall at constant velocity to the bottom of the pond.

(i) When the stone enters the water, what two forces act on it?

....

.... **[2]**

(ii) What name is give to the constant velocity of the stone as it nears the bottom of the pond?

....

.... **[1]**

[Total: 12]

Model Test Paper II
PHYSICS

M.M: 90   **Time:** 1 hour 15 min

1 (a) The diagram below shows a child on a park swing.

(i) On the diagram above, state in which position **A**, **B**, **C** or **D** would the air resistance on the child be a maximum value.

.... **[1]**

(ii) Discuss the energy changes taking place as the child oscillates freely and comes to rest after some time.
(One mark for correct spelling, punctuation and grammar.)

....

....

....

.... **[1+1]**

(iii) The rubber seat of the swing hangs from two sets of metal chains. Explain why the metal chains feel cold when the child grips them with both hands, but the rubber seat does not.

....

....

.... **[2]**

(b) The diagram below shows the main losses of thermal energy from a house.

30% through roof

12% through windows

28% through cavity walls

15% through gaps and cracks around windows and doors

loss through floor

(i) Determine the percentage of heat loss from the floor of the house.

loss = ... % **[1]**

(ii) How can the heat loss through the flooring be reduced?

.... **[1]**

(iii) The windows in the house are double-glazed.
The diagram below shows a section
through a double-glazed window.

trapped air

panes of glass

Discuss how the heat loss through the double-glazed window is reduced compared to a window that has a single pane of glass.

....

....

.... **[2]**

(iv) Explain why fitting shiny aluminium foil behind the radiators in the house would reduce heat loss.

....

.... **[1]**

[Total: 11]

2 An aeroplane is about to take off. On the ground, its potential energy is zero and it has a fully loaded mass of 60,000 kg. When it is about to leave the ground, it has a velocity of 144 km per hour.

(a) Express the initial velocity of plane in metres per second.

....

.... **[1]**

(b) Calculate the kinetic energy of the plane, when it is about to leave the ground.

....

.... **[2]**

(c) Calculate the potential energy of the plane, when it reaches its normal cruising altitude of 9,000 m. (Take 'g', the value of the acceleration due to gravity = 10 N/kg.)

....

.... **[2]**

(d) When the plane lands at its landing speed of 144 km per hour, it decelerates uniformly for 15 seconds until it reaches the speed of 36 km per hour. It maintains this speed for 30 seconds before it decelerates uniformly to rest in 15 seconds.

(i) Draw a velocity time graph for the plane's landing journey, from touchdown until it stops.

[5]

(ii) Determine the deceleration during the first 15 senconds after landing.

....

....

.... [2]

(iii) Determine the total distance travelled by the plane after landing.

....

....

.... [2]

[Total: 14]

3 **(a)** A lady stands 5 metres away from a plane mirror. If she now walks 2 metres towards the mirror, what is her distance from her image?

....

....

.... [2]

(b) What property of light is exhibited in each of the following situations:

(i) Although a swimming pool may be of constant depth, the far end appears noticeably shallower than the nearer end.

....

.... [1]

(ii) A prism breaks up white light into its constituent colours.

....

.... [1]

(c) When the time base is set to 5μs/cm on a cathode ray oscilloscope, 3 waves from a signal ocupy 12cm. Calculate the frequency with which the waves are being generated.

....

....

.... **[3]**

(d) To locate the submarine, a ship emits sound waves vertically downwards. The speed of sound in water is approximately 1500 m/s.

If the ship receives an echo after 0.2 seconds, how deep is the submarine?

....

.... **[1]**

If the submarine now dives a further 350 m, how long will it take for an echo to return to the ship?

....

.... **[1]**

(e) Given that the critical angle for glass is 42°, draw the path taken by a ray of light in each of the glass prisms shown below:

[1]

(f) Rank the following electromagnetic waves in order of decreasing wavelength:

Yellow light, radio waves, X-rays, microwaves, blue light, infrared waves.

....

.... **[2]**

[Total: 14]

4 (a) A rubber balloon acquires a positive charge when it is rubbed by a piece of cloth.

   (i) In terms of electrons, explain how the balloon acquires a positive charge.

....

.... **[2]**

(ii) The diagram below shows two positively charged balloons which
are placed alongside each other.

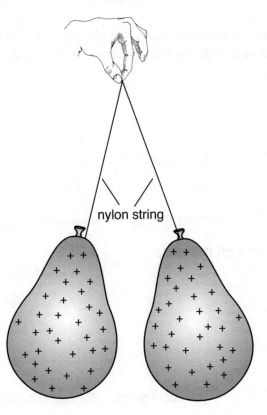

nylon string

On the diagram above, using arrows show the size and direction
of the electric force experienced by each of the balloons. **[2]**

(b) An electric kettle has 350 g of water at 20 °C. The kettle is switched on. The
current in the heating element of an electric kettle operating at 230 V is 7.2 A.

Calculate:

(i) the resistance of the heating element of the heater,

resistance = ... Ω **[3]**

(ii) the electrical power of the kettle,

power = ... W   [3]

(iii) how long it takes for the water to reach its boiling point (100 °C).
It takes 4200 J of heat to change the temperature of 1 kg of water
by 1 °C,

time = ... s   [4]

(iv) the charge flowing through the kettle.

charge = ... unit: ....   [4]

[Total: 18]

5   (a) Explain what is meant by the *half-life* of a radioactive substance.

....

....   [2]

(b) The table below shows the half-lives of three emitters of β particles.

β particle emitters	half-life
strontium-90 ($^{90}_{38}$Sr)	28 years
lead-214 ($^{214}_{82}$Pb)	27 minutes
carbon-14 ($^{14}_{6}$C)	5700 years

(i)   Which part of the atom does the β particle come from?

.... **[1]**

(ii)   State **two** properties of β particles.

....

.... **[2]**

(iii)   β particles are very dangerous to human beings. Explain why they
            are dangerous and state how you can minimise exposure from
            such a radiation.

....

....

.... **[2]**

(iv)   Samples of the three β particle emitters have the same number of atoms.
            Which sample has the highest activity? Explain your answer.

....

....

.... **[2]**

(v)   What fraction of the nuclei of lead-214 has **decayed** after a
            time of 81 minutes?

fraction = ... **[3]**

[Total: 12]

6 (a) Explain why metals are good conductors of heat.
        (One mark for a correctly ordered answer.)

....

....

....

....

.... [3+1]

(b) The gas atoms around us are constantly colliding with each other. The diagram below shows a helium atom making a head-on collision with a stationary carbon atom.

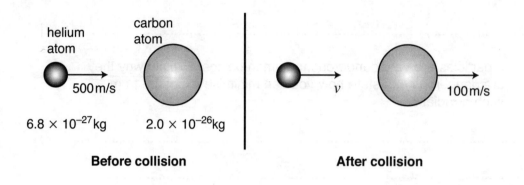

Before collision   **After collision**

(i) Calculate the initial momentum of the helium atom.

momentum = ... kg m/s **[3]**

(ii) Use the information provided on the diagram above to determine the velocity of the helium atom after its collision with the carbon atom.

velocity = ... m/s **[4]**

[Total: 11]

7 (a) All logic gates use digital signals. With the aid of a sketch, explain what is meant by a digital signal.

....

.... [2]

(b) The diagram below shows a logic gate.

(i) Name the logic gate shown above.

.... [1]

(ii) Complete the truth table for the logic gate.   **[2]**

A	B	output
0	0	
0	1	
1	0	
1	1	

(c) The diagram below shows a potential divider circuit consisting of a light-dependent resistor (LDR).

(i) State how the resistance of the LDR is affected by the intensity of light.

....

.... **[1]**

(ii) In sunlight, the resistance of the LDR is 1200 Ω. Calculate the voltage across the LDR.

voltage = ... V **[3]**

(iii) State whether or not the light-emitting diode (led) is lit.

.... **[1]**

[Total: 10]

Model Test Answers: Paper 1

Question	Answer	Mark

1 a Renewable sources of enery are sources that can be replaced so that their overall supply is not depleted.   **1**

b Wind, tidal, solar.   **1+1+1**

c Advantage of wind power: Several windmill units can be linked to provide a powerful source of energy   **1**

Disadvantage: Erratic; it will not generate any power on calm days.   **1**

[*For other sources of energy, one reasonable advantage and one reasonable disadvantage should be stated.*]

d i Chemical energy changes to gravitational potential   **1**

ii gravitational potential changes to kinetic   **1**

iii kinetic changes to heat (and possibly some sound).   **1**

e i Radiation
ii convection   **1+1**

2 a There is a reaction from the board vertically upwards.   **1**

Examiner's tip:
The swimmer is standing still; therefore the net force must be zero. You already know that the weight acts downwards hence the reaction must be equal but opposite to the weight.

b weight = mass × gravitational field strength ($W = mg$)   **1**
weight = 62 × 10 = 620 (N)   **1**

c i Clockwise moment   **1**

ii The force F is greater than the weight   **1**
because the distance of the force F from the pivot **X** is smaller.   **1**

Examiner's tip:
moment = force × distance from pivot
The clockwise moment due to the weight is equal to the anticlockwise moment due to the force F. Since the distance of the force F from the pivot is smaller than the distance of the weight from the pivot, the force F must be larger than the weight.

d i The swimmer has a <u>constant</u> acceleration up to 1.5 s.   **1+1**
After 1.5 s the swimmer is in the water and decelerates.   **1**
Any further detail.   **1**
(E.g.: The deceleration is not constant or the swimmer slows down because of resistance from the water)

Examiner's tip:
There are two marks for the first statement.
For writing down '*the swimmer accelerates*' you will only get one mark.

ii acceleration = rate of change of velocity or acceleration = gradient of graph   **1**
$a = \dfrac{10-0}{1.0}$ **1**
$a = 10$ (m/s^2)   **1**

Examiner's tip:
Always show your working. If you do the question by calculating the gradient from the velocity against time graph, then show the 'triangle used' on the actual graph. Examiners would expect working here since the acceleration for free fall is known by most candidates.

3 a The oscillations are at right angles to the wave direction.   **1**

b i The speed of light decreases as it enters the plastic.   **1**

ii

Ray internally reflected at the first interface.   **1**
Ray internally reflected at second surface and emerging parallel to the original ray.   **1**

c The sound is diffracted at the open door.   **1**
Light is not diffracted at the door.   **1**
Any further detail.   **1**
(E.g.: The sound from the piano has a wavelength similar to the width of the door or the wavelength of light is too small to be diffracted.)

Examiner's tip:
The last mark is definitely for an A* grade candidate.

d i speed = frequency × wavelength ($v = f\lambda$)   **1**
$\lambda = \dfrac{v}{f} = \dfrac{340}{220}$ **1**

wavelength = 1.54 m ≈ 1.5 m   **1**

ii The sound from both loudspeakers interferes destructively.   **2**
(Allow one mark for writing '*waves cancel each other*' or '*peaks and troughs combine*'.)

4 a

Both rays shown to diverge at the lens.　**1**
Correct distance marked for the focal length.　**1**

b i　An image that cannot be formed on a screen. **1**

Examiner's tip:
Resist writing '*virtual image is not real*', unless you aim to explain what is meant by a real image.

ii

Correct path for one ray.　**1**
Correct path for second ray.　**1**
Image located correctly and labelled **I**.　**1**

iii　The image is magnified　**1**
and upright.　**1**

5 a i　voltage across 8Ω resistor = 16V　**3**

ii　current through 6Ω resistor = 1.33A　**3**

b　the kilowatt-hour (kWh)　**1+1**

c i　3.7 units　**3**

ii　$0.37　**3**

6 a　The current in the coil **A** creates a magnetic field in the soft-iron rod.　**1**
This magnetic field is changing when the switch is closed.　**1**
This changing magnetic field links the coil **B** and hence a voltage is produced in coil **B**.　**1**
Eventually, the constant current produces a constant magnetic field. There is no change in the magnetic field, hence there is no voltage produced in coil **B**.　**1**

Examiner's tip:
This is a tough question. It is testing whether or not you understand how a 'transformer' works.
Learn this section well.

b i

$$\frac{\text{voltage across primary}}{\text{voltage across secondary}} = \frac{\text{turns on primary}}{\text{turns on secondary}}　\textbf{1}$$

$230 \div 3.8 = 5200 \div$ turns on secondary　**1**

turns on secondary $= \dfrac{5200}{230} \times 3.8 = 86$　**1**

Examiner's tip:
Candidates often make a mess of rearranging this equation. You know that it is a step-down transformer. Hence the number of turns on the secondary will be smaller than that on the primary coil by a factor of

$$\frac{5200}{60.5} = 60.5$$

Therefore, the number of turns on the secondary coil

must be $\dfrac{5200}{60.5} = 86$

ii　input power = output power　**1**
$230 \times 0.012 = 3.8 \times$ current　**1**

current $= \dfrac{230 \times 0.012}{3.8} = 0.73\,(A)$　**1**

Examiner's tip:
In a step-down transformer, the voltage is stepped down but the current increases. The current increases by the same factor as the ratio of the number of turns.
Hence current in the secondary coil
$= 12\,\text{mA} \times 60.5 = 730\,\text{mA} = 0.73\,\text{A}$

c　High voltage means smaller current therefore less energy is lost in the overhead cables as heat.　**1**

7 a　An atom of an element, the nucleus of which has the same number of protons　**1**
but different number of neutrons.　**1**

b　Background radiation is radiation produced by external sources.　**1**
Rocks (granite) or outer space or nuclear fall out or medical sources.　**1**

c i　protons = 6　**1**
neutrons (= 14 − 6) = 8　**1**

ii　90 ÷ 22.5 = 4　**1**
number of half-lives = 2　**1**
age = 5600 × 2 = 11 200 (years)　**1**

Examiner's tip:
The question is definitely something to do with half-lives. You can start with 90 counts per hour and keep dividing the answer by 2 and stop when you reach 22.5 count per hour. The number of times you have to do this will give you the number of half-lives. Therefore:
$90 \div 2 = 45 \;\rightarrow\; 45 \div 2 = 22.5$
one half-life　　　another half-life

8 **a**

b   Upward force at P: 120g N
(1200N, taking g = 10 N/kg)   **1**

     Upward force at Q: 80g N   (800 N)   **1**

c   New force at P: 70g N   (700 N)   **1**

     New force at Q: 130g N   (1300 N)   s   **1**

d   Pressure exerted by man = Force ÷ Area =
$800N \div 80cm^2 = 10\ N/cm^2$ **1**

     Pressure in Pa = $800\ N \div 80 \times 10^{-4}m^2 =$
$10 \times 10^4 = 10^5\ Pa$ **1**

e   **i**   When the stone enters the water, the
gravitational pull of the earth (weight)
and the resistance of the water (upthrust)
act on it   **1+1**

    ii   Terminal velocity   **1**

QWC mark: One mark is awarded if answers are
presented in logical manner and text is legible.

Model Test Answers: Paper 2

Question Answer	Mark	Question Answer	Mark

1 a i Position **C**. 1

Examiner's tip:
The air resistance depends on the speed of the object. The drag force increases as the speed of the object increases. The question wants you to work out in which position the speed of the child is a maximum. The speed of the child is a maximum at position **C**.

ii *Any 2 from:*
Gravitational potential energy to kinetic energy from **A** to **C**.
Kinetic energy to gravitational potential energy from **C** to **D**.
Kinetic energy (or potential energy) to heat as the oscillations die out. **1+1**
QWC mark: The mark is awarded for sentences without error in spelling, punctuation and grammar. 1

iii Metal is a good conductor of heat. 1
Heat flows from the hands to the chains, so the hands feel cold. 1

b i 15% 1

ii Loss through the flooring can be reduced by having a thicker carpet or an underlay. 1

iii Loss of heat by conduction is reduced because air is a poor conductor of heat. 1

iv The shiny aluminium reflects radiation back into the house. 1

2 a 40 m/s

b 48 MJ

c 5.4 GJ

d i 5

ii deceleration = 2 m/s² 2

iii total distance = area under graph = 750 m 2

3 a 6m 2

b i refraction **ii** dispersion **1+1**

c 50kHz 3

d 150m, 0.667 seconds **1+1**

e 3

1 mark for each ray diagram adds up to a total of 3 marks.

f radio waves, microwaves, infrared waves, yellow light, X-rays 2

4 a i The rubbing action removes electrons from the balloon. 1
Electrons have a negative charge. The balloon is left with a net positive charge. 1

ii

Two arrows in opposite directions. 1
Equal sized arrows (showing the same force experienced by each balloon) 1

b i resistance = voltage ÷ current ($R = \dfrac{V}{I}$) 1

$R = \dfrac{230}{7.2}$ 1

$R = 31.9\,(\Omega) \approx 32\,(\Omega)$ 1

ii power = voltage × current ($P = VI$) 1
$P = 230 \times 7.2$ 1
$P = 1700\,(W)$ 1

iii energy needed to change the temperature of water by 1°C = $\dfrac{350}{1000} \times 4200$ 1

energy needed to boil the water

$= \dfrac{350}{1000} \times 4200 \times (100 - 20)$ 1

energy needed to boil the water
$= 1.18 \times 10^5\,J$ 1

time = energy ÷ power = $\dfrac{1.18 \times 10^5}{1700}$ 1

time = 69 (s)

iv charge = current × time ($Q = It$) 1
charge = 7.2×69 1
charge = 500 1
unit: coulombs or C 1

Examiner's tip:
You can always work out what the unit for charge from the equation $Q = It$. The unit for charge is:

ampere $\times$ second $\rightarrow$ As

If instead of the 'coulomb' you write down either 'ampere second' or 'As', then the examiner has to give you full credit for the correct physics.

5 a The half-life of a substance is the average time taken for half **1**
the number active nuclei to decay. **1**

Examiner's tip:
Quite often candidates will know that this definition involves something being halved. You have the choice of stating either '*number of active nuclei*' or the '*activity*'. Reference to the nuclei is important. It would be wrong to state that '*half-life is the time taken for the substance to halve its mass*'.

b i The nucleus. **1**

Examiner's tip:
Do not simply state that 'atoms' release β particles. It is very important to appreciate that radioactivity is to do with the changes taking place within the nuclei. Some candidates wrongly assume that the β particles are the orbiting electrons within the atoms.

ii *Any two from:* **1+1**
They are electrons.
Carry a negative charge.
Can be stopped by a thin sheet of aluminium.

iii They destroy living cells. **1**
Use a thin sheet of aluminium to shield from the source. **1**

iv The lead sample will have the largest activity. **1**
The nuclei decay in a short period of time (compared with the other samples). **1**

v 81 minutes = 3 half-lives **1**

fraction of nuclei left after 81 minutes =

$$\frac{1}{2} \times \frac{1}{2} \times \frac{1}{2} = \frac{1}{8}$$ **1**

fraction of nuclei that have decayed =

$$1 - \frac{1}{8} = \frac{7}{8}$$ **1**

Examiner's tip:
The question requires careful reading because it does not want you to just determine the fraction of nuclei **left** within the sample after 81 minutes. A mathematical way of getting the answer would be to use:

number of nuclei decayed = $1 - (\frac{1}{2})^n$,

where n = number of half-lives.

6 a Metals are good conductors because of free electrons. **1**

These electrons can easily diffuse through the metal. **1**

Electrons gain thermal energy by colliding with the vibrating atoms at the hot end of the metal and transport this energy to the cooler atoms at the cold end. **1**

QWC mark: The mark is awarded for statements made in a logical order. **1**

b i momentum = mass $\times$ velocity **1**
momentum = $6.8 \times 10^{-27} \times 500$ **1**
momentum = $3.4 \times 10^{-24} (\text{kg m/s})$ **1**

Examiner's tip:
Momentum is a vector quantity. It is therefore very important that you use 'velocity' in this equation and not 'speed'. Use of the term speed in the equation will lose you the first mark.

ii initial momentum = final momentum **1**
$3.4 \times 10^{-24} = (6.8 \times 10^{-27} \times v) +$
$(2.0 \times 10^{-26} \times 100)$ **1**
$v = (3.4 \times 10^{-24} - 2.0 \times 10^{-24}) \div$
6.8×10^{-27} **1**
$v = 206 (\text{m/s})$ **1**

Examiner's tip:
The key to doing this type of question is to set up an equation using the law of conservation of momentum and then solve for the unknown velocity v.

7 a

Correct sketch for a digital signal (see the
sketch above).   **1**
A digital signal has only two values or
levels of voltage.   **1**

Examiner's tip:
You can also state that a digital signal only has 'high' or
'low' values.

b i AND gate

ii

A	B	output
0	0	0
0	1	0
1	0	0
1	1	1

2

Half mark is awarded for each correct entry in
the truth table $-4 \times (\frac{1}{2}) = 2$

c i

The resistance of the LDR decreases as the
intensity of light increases.   **1**

ii total resistance $= 1200 + 600 = 1800\,\Omega$ **1**

current $= \dfrac{9.0}{1800} = 0.005\,\text{A}$ **1**

voltage $= IR = 0.005 \times 1200 = 6.0\,\text{V}$ **1**

Examiner's tip:
The current in a series circuit is the same. Use this idea to
calculate the voltage across the LDR and do not forget to
use the resistance of the LDR and not the resistor. Using
$600\,\Omega$ would give the wrong answer. If however, you show
all your working, then you would only lose the final mark.

iii The led is not lit.   **1**

Examiner's tip:
The NOT gate inverts the input. The input voltage is 6.0 V,
which is closer to the 9.0 V than the 0 V. Therefore the input
to the gate is 'high'. The NOT gate changes this high input
into a low output. Consequently, the light-emitting diode is
not lit.